Power Systems

For further volumes:
http://www.springer.com/series/4622

Liisa Haarla · Mikko Koskinen ·
Ritva Hirvonen · Pierre-Etienne Labeau

Transmission Grid Security

A PSA Approach

 Springer

Prof. Liisa Haarla
Department of Electrical Engineering
Aalto University School of Electrical
    Engineering
PO Box 13 000
00076 Aalto
Finland
e-mail: liisa.haarla@aalto.fi

Mikko Koskinen
Fingrid Oyj
PO Box 530 23 B
00101 Helsinki
Finland
e-mail: mikko.koskinen@fingrid.fi

Dr. Ritva Hirvonen
Fingrid Oyj
PO Box 530
00101 Helsinki
Finland
e-mail: ritva.hirvonen@fingrid.fi

Prof. Pierre-Etienne Labeau
Service de Métrologie Nucléaire
Université Libre de Bruxelles
avenue F. D. Roosevelt 50
1050 Bruxelles
Belgium
e-mail: pelabeau@ulb.ac.be

ISSN 1612-1287

e-ISSN 1860-4676

ISBN 978-1-4471-2628-7

ISBN 978-0-85729-145-5 (eBook)

DOI 10.1007/978-0-85729-145-5

Springer London Dordrecht Heidelberg New York

British Library Cataloguing in Publication Data
A catalogue record for this book is available from the British Library

© Springer-Verlag London Limited 2011
Softcover reprint of the hardcover 1st edition 2011

Apart from any fair dealing for the purposes of research or private study, or criticism or review, as permitted under the Copyright, Designs and Patents Act 1988, this publication may only be reproduced, stored or transmitted, in any form or by any means, with the prior permission in writing of the publishers, or in the case of reprographic reproduction in accordance with the terms of licenses issued by the Copyright Licensing Agency. Enquiries concerning reproduction outside those terms should be sent to the publishers.

The use of general descriptive names, registered names, trademarks, etc. in this publication does not imply, even in the absence of a specific statement, that such names are exempt from the relevant protective laws and regulations and therefore free for general use.

The publisher makes no representation, express or implied, with regard to the accuracy of the information contained in this book and cannot accept any legal responsibility or liability for any errors or omissions that may be made.

Cover design: eStudio Calamar, Berlin/Figueres

Printed on acid-free paper

Springer is part of Springer Science+Business Media (www.springer.com)

Preface

This book presents a method of adapting the traditional probabilistic safety analysis (PSA) to the security analysis of a power system. The book proposes are liability model for a power system, where the possible failures of substation operations after grid faults are analysed and their impact on the power system dynamics and security is taken into account. The book presents the method in such a way that the reader is guided and equipped to build up a similar model. This reliability approach is suitable and applicable to real transmission grids, which can have hundreds of substations and lines.

In the reliability analysis of transmission systems, the basic phenomena and reliability problems are at the system level. The analysis of the local perspective, for example the outages of single components, is not adequate to capture the whole picture. So far, there have not been systematic methods that would combine local and system level issues in a similar way as this book presents. The method adapts the traditional probabilistic safety analysis (PSA) to the security analysis of a power system and includes the simulation of grid dynamics after grid faults and component failures in the modelling. This combination of different (traditional) tools enables a systematic security analysis where the connection of the failures ofsingle devices and the system level consequences, for example a system breakdown, is possible.

The approach and methods presented in the book are based on the dissertation of one of the authors, Professor Liisa Haarla (formerly Pottonen). Her background is in the system planning, reliability and protection of power systems. Usually, only a few readers have deep enough knowledge to understand the insights of dissertations and few people actually read the dissertation. This book presents the basic ideas of the dissertation, but in addition to that, it provides a lot of background information and in doing so, presents the idea in a larger context. This helps the readers to understand the applied methods better.

The book also gives theoretical knowledge about reliability engineering and power systems. Chapter 3 presents the basic concepts of transmission grid planning. The main author of this chapter is Senior Specialist Mikko Koskinen, who has worked for over 20 years in that field. The reader is given a deeper appreciation of the dynamics and reliability aspects of a meshed transmission grid. The book also presents the basic concepts of reliability theory, written by Professor Pierre-Etienne Labeau, whose expertise is in reliability engineering and PSA methodology for nuclear applications. This chapter helps the reader to understand the essential features of the proposed method. Doctor Ritva Hirvonen introduces the electricity market design, renewable generation, and emerging technologies that affect transmission system and its reliability. Her background is in operational and system planning of the network and in electricity markets. The book provides references to the literature that the interested reader can use for deepening his/her knowledge in the various domains combined in this work.

The book is practical minded, it concentrates on issues encountered when an analysis of this kind has to be completed for an existing transmission grid. This perspective is quite rare in books, because experts making practical security analyses for a real power system are seldom interested or have time to write books.

Acknowledgments

The authors are greatly indebted to late Research Professor Urho Pulkkinen, who originally introduced the idea of using the PSA approach to study the reliability of power systems. The book developed this idea into practical application in the real-size transmission grid.

Fingrid Oyj, Transmission System Operator in Finland, has provided the valuable fault and failure statistics of the transmission grid for analysis. The load flow and dynamic model of Fingrid Oyj's grid has been applied to study the feasibility of the method. These contributions from Fingrid Oyj, enabled the analysis with a real network.

The authors thank Mr. Ilkka Luukkonen for designing and drawing most of the figures presented in the book.

Contents

Abbreviations and Symbols

A	Availability
A	Event
H	Set of Events
BFR	Breaker Failure Relay
CB	Circuit Breaker
CIGRE	International Council on Large Electric Systems
CT	Current Transformer
D	Differential Relay
E	Voltage
ENTSO-E	European Network of Transmission System Operators
ET	Event Tree
f	Frequency
$f(t)$	Failure Density (As a Function of Time)
FMEA	Failure Mode and Effect Analysis
FT	Fault Tree
FV	Fussell-Vesely's Measure of Importance
HVDC	High Voltage Direct Current
IEC	International Electrotechnical Commission
IEEE	Institute of Electrical and Electronics Engineers
LE	Line End
LOLP	Loss of Load Probability
MCB	Miniature Circuit Breaker
MCS	Minimal Cut Set
MTTF	Mean Time to Failure
MTTR	Mean Time to Repair
$N-1$	The $N-1$ criterion is a method of providing reliability to systems. According tothis criterion, the system is sufficiently reliable if it is able to operate under any unplanned outage of a component due to a single cause. In a power system, the criterion means that the loss of any line, busbar, generator or transformer after a single power system

	fault will not cause overloading of the remaining components or other problems
NERC	North American Reliability Council
Nordel	Nordel was the organisation for co-operation between the transmission system operators the Nordic countries, i.e. Denmark, Finland, Iceland, Norway and Sweden until 2009 when ENTSO-E started its operation
NPP	Nuclear Power Plant
p	Probability
P	Power (Active Power)
PLC	Power Line Carrier (Telecommunication)
POTT	Permissive Overreach Transfer Trip
PSA	Probabilistic Safety Analysis
Q	Reactive Power
q	Unavailability
RAW	Risk Achievement Worth
RDF	Risk Decrease Factor
RIF	Risk Increase Factor
RRW	Risk Reduction Worth
$R(t)$	Reliability (As a Function of Time)
SAIDI	System Average Interruption Duration Index
SAIFI	System Average Interruption Frequency Index
SB	System Breakdown
SF6	Sulfur Hexafluoride
Tele	Telecommunication Channel
T_{DT}	Mean Downtime
T_{TF}	Mean Time to Failure
T_{TR}	Mean Time to Repair
TSO	Transmission System Operator
U	Voltage
VT	Voltage Transformer
W1	Busbar 1
W2	Busbar 2
X	Reactance
yr^{-1}	Per Year
Z	Distance Relay
Z STA	Static Distance Relay
δ	Rotor Angle, Angle Difference between Voltages
λ	Failure Rate
μ	Repair Rate

Chapter 1
Introduction

The basic functions of our modern society are dependent on electricity supply. Lighting, water supply, heating, traffic and fuel supply, food storage and delivery, payment transactions, telecommunication, and information systems are heavily affected by interruptions. Therefore, the electricity system can be regarded today as the most important infrastructure. Blackouts especially are likely to induce significant economic consequences, both direct and indirect. Costs of a one-day blackout could be about 0.5% of the gross domestic product excluding social consequences such as deaths and injuries [1]. With long-lasting system breakdowns, it is relevant to evaluate the risks for society rather than just the costs for individual customers.

The restructuration carried out in the electric energy industry to enable competition introduced the unbundling of vertically integrated energy companies, within which such functions as generation, supply, transmission, and distribution commonly resided together before. Thus, it was evident that the responsibility for power system security, such as the continuity and quality of electricity supply, was within this type of company. In electricity markets, generation and supply of electricity are subject to competition whereas the transmission and distribution of electricity are still monopolies by nature. In electricity markets, the transmission and distribution companies (Transmission System Operator and Distribution System Operator) have responsibility for the continuity, reliability, and quality of electricity supply to the customers. This liability has been given by the legal framework, which defines the roles and duties of different participants acting in the electricity market.

The introduction of electricity markets has brought economic aspects and cost awareness more in focus. The importance of the power system and the requirement to be effective have forced the system operators to analyze more in detail the benefits and costs of investments and maintenance in the power system. Furthermore, the effects of unreliable electricity supply on modern society have been under consideration. Kariyki and Allan [2] present an example of the methods developed to evaluate the costs of energy not served for different consumer types. These customer outage costs include the economic losses due to outages, not just

L. Haarla et al., *Transmission Grid Security*, Power Systems,
DOI: 10.1007/978-0-85729-145-5_1, © Springer-Verlag London Limited 2011

the cost of energy not supplied. In some countries, the electricity distributors have to pay compensation to customers if electricity was not delivered [3].

By far, there has been little analytic or simulation work on the transmission system security. The reasons are the complexity of the problem, the difficulty of creating proper models and getting relevant data for the models. The method presented in this book is an approach to treat the security systematically using an established reliability method, probabilistic safety analysis. The method suits best for the power systems where the dynamic behavior after faults plays a role and the stability has to be considered.

The mechanisms that lead to a power system breakdown are diverse in different parts of the transmission grid. Thus, quantitative analysis is needed for estimating the contribution of different fault locations and different grid components to the system breakdown.

References

1. Kirschen DS (2002) Power system security. Power Eng Journ 16(5):241–248
2. Kariyki KK, Allan RN (1996) Evaluation of reliability worth and value of lost load. IEE Proc Gener Transm Distrib 143(2):171–180
3. CEER (2008) 4th benchmarking report on quality of electricity supply. Ref: C08-EQS-24-04. http://www.energy-regulators.eu/portal/page/portal/EER_HOME/EER_PUBLICATIONS/CEER_ERGEG_PAPERS/Electricity/2008. Accessed 13 May 2010

Chapter 2
Grid Security: Problem Statement

2.1 System Characteristics Affecting Reliability Analysis

2.1.1 A Meshed Grid Configuration

In a way, the meshed structure of the grid is a consequence of reliability requirements, as actually a meshed system can be seen as a specific application of the redundancy principle. Often the main transmission grid is meshed whilst the local distribution grid is radial. The consequence of a component failure in the radial system is straightforward: all customers downstream the faulted feeder will face an interruption. In a meshed system, a failed component usually does not lead to any interruptions on connecting customers, because the electricity flow always has parallel paths. This is a direct consequence of the common principle applied for transmission grids: the $N - 1$ principle. This means that the system always has to withstand the loss of one component, in all operational situations including also planned maintenance outages. This redundancy is because it would be intolerable to have wide-area interruptions too frequently. Secondly, because electricity storage in a large scale does not yet exist, the power flow must continue in spite of single faults. The $N - 1$ criterion is usually applied for primary components, for example transmission lines, generators and transformers. In addition to surviving the loss of a single component, the system should cope with the dynamic fault sequence caused by the fault and the consequential loss of the component. In practice, some of those faults may be drastic for the system. For example, short circuits with large currents can jeopardize the system stability. Hence, dynamic simulations help to identify the consequences of these faults.

Even though a meshed system offers several routes, it is not always self-evident that the system would survive the transition into a post-fault state and that continuous operation at this stage would be possible. A power system is a dynamic system where generators with big rotating masses have inertia. Transients, such as short circuits, start electromechanical oscillations between different generators or generator groups. The fault duration, location, and power flow define if the

L. Haarla et al., *Transmission Grid Security*, Power Systems,
DOI: 10.1007/978-0-85729-145-5_2, © Springer-Verlag London Limited 2011

transition to the post-fault state after a fault is stable. In addition, it is necessary to ensure if the transmission capacity of the post-fault system is adequate, since lines and other grid components have a limited transmission capacity.

In a system reliability analysis, it is, therefore, crucial to analyse the dynamic performance of the system immediately after faults, i.e. to perform a stability analysis. It is not enough just to check that the steady-state post-fault situation is acceptable. If the transition is unstable, the system may collapse due to transient instability or electromechanical oscillations and the system never reaches any post-fault steady-state power flow but ends up to a system breakdown.

In a radial system, it is usually enough to check the continuity of the transmission path, but no stability analyses are needed. In the short-term future, with distributed generators, the situation might change.

Grid topology changes can be made only at locations where circuit breakers exist. Most circuit breakers are located at substations, which are nodes of the network. In addition, some important functions for the reliable operation of the system, for example telecontrol terminal equipment and relay protection systems, are located at substations. The chosen substation scheme also defines the possible switching actions, topology changes, and the availability of a feeder.

Figure 2.1a and b presents two different substation arrangements whilst in Fig. 2.2 different breaker schemes are shown. In a double busbar substation with a two-breaker arrangement (Fig. 2.1a), each feeder has two circuit breakers. This ensures the high availability of a single feeder since one breaker is sufficient for the operation of the feeder. The two-breaker system offers flexible possibilities to connect the feeders into any of two main busbars. The two-breaker system is not vulnerable in the case of a busbar fault. Indeed, operation can continue through the other main busbar after the protection has tripped the faulted busbar. The main drawback of the two-breaker system is that it is more expensive in comparison with other alternatives.

The single busbar substation with a sectionalizing circuit breaker (Fig. 2.1b) can be a viable solution in a meshed sub-transmission network where all feeders are supplied from both directions and the transmission function is not so crucial.

Fig. 2.1 Two different substation arrangements with four feeders. Black squares represent circuit breakers **a** presents a double busbar substation with two circuit breakers per one feeder, and **b** presents a single busbar substation with a sectionalizing circuit breaker

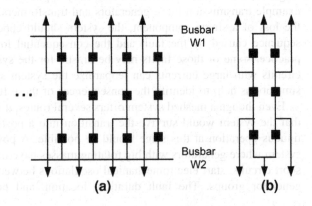

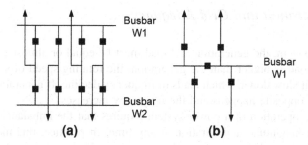

Fig. 2.2 Two different circuit breaker arrangements, where black squares represent circuit breakers **a** presents a *1½-breaker arrangement*, which is a substation scheme, where all circuit breakers are usually closed in normal operation. Since each feeder is fed from two sides, all feeders are in operation also after a busbar fault **b** presents a *ring substation*, which is a single busbar substation, where the busbar is a closed loop with circuit breakers in series. The ring busbar arrangement allows an uninterrupted operation of all feeders

The sectionalizing breaker gives some advantage in the case of a busbar fault and during the scheduled maintenance of the equipment directly connected to the main busbar. However, the grouping of feeders is fixed and the availability of radial feeders is relatively low. This solution is more at home at gas-insulated switchgear (GIS) where the reliability of single components is good and the number of gas-tight pressure vessels is reasonably low.

The one-and-a-half circuit breaker arrangement (Fig. 2.2a) falls somewhere between the two-breaker and one-breaker systems as regards the availability and the cost per feeder. The availability of a single feeder is close to that of the two-breaker system due to two-way infeed. The grouping of feeders is not as flexible as in the two-breaker system. The ring substation in Fig. 2.2b can be the first step towards the formation of a 1½-circuit breaker system. With six feeders, the 1½-circuit breaker configuration is actually a ring substation.

In reliability evaluation, it is important to distinguish the aspects linked to *grid customers* (generators and electricity consumers) and the *operability* of the grid, both described by performance indicators. What happens in the grid is not important as such if only the grid remains operational and the customers get the service they need. If there are outages of components in a meshed grid, but the customers get their service, the customer-related performance indicators remain good. For a meshed transmission grid, the relevant performance indicator expresses the occurrence of outages for a wide area. In practice, this can happen after a major disturbance, cascading outages or a system breakdown.

There is a difference between the reliability challenges for radial and meshed grids. In a radial system, an outage always leads to an interruption to some customers but not to all of them. In a meshed system, an outage does not lead to problems for a single customer, but the consequences of a system breakdown may affect a large number of customers spread across the control zones of transmission system operators. These differences mean that reliability analyses of meshed and radial systems are inherently different.

2.1.2 Generation and Grid Adequacy

In a power system, the generation and load must be equal or almost equal at any time. If the load is bigger than the generation, the rotating speed of synchronous machines will slow down, which leads to frequency decline. If generation exceeds the load, the opposite happens and the frequency increases.

The stable operation of a power system requires that the imbalance should be within given magnitude and duration at any time. In practice, this means that a sufficient balance shall be maintained at any moment. The rotating machines cannot endure the operation with a high over-speed because of the too large mechanical stresses they induce. Under-frequency is dangerous due to the risk of mechanical resonances. Therefore, the generation (and load) needs to be controlled.

Usually the generation and the consumption do not locate near each other since other issues define their optimal or possible location. In order to feed the consumption, enough transmission capacity is needed. Adequate transmission capacity involves especially sufficient thermal loading capability and high enough transmission voltages for the transmission distance and transmitted power.

Adequacy evaluation is twofold: both generation and transmission grid adequacy must be considered.

2.1.3 Impacts of System Dynamics

During over-frequency situations, the operators can always reduce the generated power or even trip generators. Under-frequency calls for reserve generation, else some load must be shed; otherwise the system may collapse. In a large synchronous system consisting of several interconnected parallel operating subsystems, regional surplus or deficit of power up to the transmission capacity of interconnections from other subsystems, however, is fully acceptable. If transmission is higher than pre-defined security limits, transmission needs to be reduced to an acceptable level. In addition to continuous load and generation variations, the transmission system is exposed to faults and disturbances that create dynamic transitions from one system state to another. The transitions are oscillatory by nature.

In order to keep the system stable, it is essential to disconnect the faults and restore the system into a state, where it can withstand a new fault (restore security). Ways to do this are for example connecting healthy components into operation rapidly, starting generation reserves or shedding load. Many line faults are temporary and high-speed automatic reclosing enables the system restoration into a secure state. The security and stability depend on the characteristics of the transmission system and generators, the magnitude of power flows, and the fault location, duration and type.

Often it is possible to get more transmission capacity if $N - 1$ security is planned in such a way that remedial operational actions (for example starting reserve power stations) are accepted after the fault and trip of the faulted component.

If this is the case, the secure operation needs three things: maintaining power flow at an acceptable level, tripping faulted components rapidly enough, and sufficient operation actions to recover the system in such a state that it can withstand a new fault. Figure 2.3 illustrates the possible chains of events that can occur if these remedial operational actions are inadequate or fail.

A delay in restoring the system after an interruption caused by a fault may lead to an overload on the remaining circuits (Fig. 2.3a) or to abnormal voltages (Fig. 2.3b). The constraint and the operational state together define the urgency of

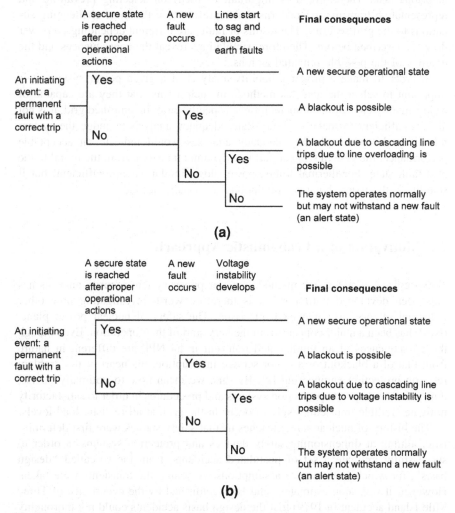

Fig. 2.3 Two simple chains of events (called event trees), which present possible consequences after a power system fault with a correct trip. A typical proper operational action is to reduce power flow by starting reserve power plants **a** thermal loading of lines defines the transmission capacity (typically short lines), and **b** system stability defines the transmission capacity (typically long lines)

the remedial actions needed. If the constraint is the thermal overloading of lines, for example, the remedial actions are dictated by the thermal time constant, which can be in the range of 10–20 min. Otherwise, the lines start to sag too much due to heat and contact with vegetation is possible. With voltage stability, the time from the initial fault occurrence to voltage collapse varies from a few seconds to several hours [1, p. 19].

The variability of operational states and transmission patterns leads to a situation, where a comprehensive reliability analysis would be a very complex and laborious task. Therefore, it is important to focus on selecting prevailing and representative transmission patterns that represent also the most challenging situations to the grid security. The results of faults are different depending on power flow, as described before. Therefore, simulations reveal the consequences and the urgency of the possible remedial actions.

In order to make a proper reliability analysis of a given power system, it is important to select the analysis methods in such a way that they are capable of capturing the relevant phenomena for each specific case. In a meshed grid, it is not always sufficient to make a steady-state adequacy analysis to judge the system reliability. Even though the adequacy analyses would indicate an acceptable performance, the system still can fail in a dynamic transition from the initial to the post-fault state. For thermal limits, power flow simulations are sufficient, but if stability is the issue, dynamic simulations are usually needed.

2.2 Motivation of a Probabilistic Approach

No specific, comprehensive methodology for power system security analysis has thus been developed until now. It is therefore worth investigating how other industrial sectors have managed this issue. The safety of nuclear power plants (NPP) has been a major concern from the very start of their operation. Even though the consequences of an unmitigated transient in an NPP are different in nature from that of a blackout or a major service interruption, the heart of the problem rests on protection engineering [2]. By this, we mean how to optimally devise, dimension, and operate protection systems and procedures in order to satisfactorily mitigate credible transients likely to occur in the grid, at all realistic load levels.

The history of nuclear safety teaches us that safety studies were first deterministic, aiming at dimensioning safety devices and protection systems in order to successfully mitigate a set of postulated accidents, from the so-called "design basis", for which conservative assumptions worsening the transients were taken. However, it was soon estimated (and later confirmed by the occurrence of Three Mile Island accident in 1979) that the design basis accidents could not thoroughly envelop all possible cases; transients outside the design basis were still likely to take place, and the corresponding residual risk had to be properly investigated in order to optimally reduce it. Probabilistic safety analyses (PSA) were then done [3] in order to complement the initial deterministic studies. Abandoning the conservative assumptions underlying the deterministic approach, PSA studies were completed on a more realistic basis, with the purpose of systematically delineating and identifying accident scenarios challenging the NPP safety. Scenarios were built at the system level, the successful or unsuccessful solicitation of protection devices generating branch points in the transient development. Each of these branch events was in parallel analysed in a top-down approach to determine which combinations of basic component failures could cause them. Probabilities were then associated to the logical decomposition of events, eventually providing occurrence frequencies associated to these scenarios. These frequencies were then interpreted as an importance weight to build up a severity ranking amongst the scenarios challenging the NPP integrity, driving the efforts of improving the plant safety towards efficient risk reduction. The probabilistic character of the analysis hence appears as the measure according to which the level of completeness of the plant environment protection must be assessed, given regulatory and economic constraints.

An analogy with power system security can quite straightforwardly be drawn. Deterministic studies can be compared with the application of the $N - 1$ criterion, through which the grid has to be proven to be correctly dimensioned (in terms of mesh structure and line capacity), i.e. to be adequate, in order to successfully face the loss of any single element, in all operational circumstances. PSA studies then appear as a natural way to complement the $N - 1$ analysis, providing a deeper insight in the potential grid vulnerabilities.

The methodology presented in this book is strongly inspired by the one that has been extensively applied to NPP safety analysis. It is, however, adapted to the peculiarities of a geographically distributed and meshed, interconnected infrastructure as a transmission network. It provides the classical 3-uples of a risk analysis, i.e. scenario identification, occurrence frequency estimation and damage assessment.

2.3   State of the Art in the Field of Power System Reliability

2.3.1   Concepts of Power System Reliability

The term *reliability*[1] is used as a general concept in power systems and in relay protection and it relates to the probability of a satisfactory operation in the long run. The reliability of a bulk power system can be measured by the quality of service the customers receive. There are different measures or indexes for reliability evaluation, for example the frequency and duration of interruptions. [4]

---

[1]   Maybe it is worth mentioning that often dependability is the general term describing reliability. IEC standard *Dependability and quality of service* uses dependability as a term of collective availability performance but points out that it is used only for general description in non-quantitative terms (IEC 60050-191).

Reliability of power systems has two aspects: adequacy and security. Power system security is the ability of the power system to withstand short circuits and other faults and sudden disturbances, for example the loss of generators or other system components. A secure power system survives after faults and disturbances (contingencies) without losing its stability or causing interruptions to customers. Thus, security relates to the robustness of the system in a context of contingencies and depends on the power system operating condition before the disturbance and the probability of disturbances. Security is a dynamic issue, it implies both the transition to the new operating point and the state of this new operating point and thus is a time varying attribute. A power system can be reliable even though it was not secure in every time instant. This is because the probability of faults is low if the system stayed insecure for short time instants [4].

Power system adequacy (discussed also in Sect. 2.1.2) is the ability of the system to supply the aggregate electric power and energy requirements of the customers at all times [4]. Adequacy deals both with generation and transmission capacity. Adequacy, being the steady-state issue, is measured by established probability-based indexes, such as the loss of load probability LOLP, the oldest index in use [5], defined the average number of days on which the peak load is expected to exceed the generation capacity [6]. Commonly used indexes are also system average interruption duration index SAIDI and system average interruption frequency index SAIFI [7]. Often these indexes are used for the performance evaluation of radial distribution systems, but they can be used for evaluating the performance of a connection point of a meshed transmission grid, too.

2.3.2 Static and Statistical Approaches

After a blackout has occurred, it can be analysed as a deterministic sequence of events, to better understand its causes and development. This does not imply that the opposite would be possible, i.e. identify via simulation all the possible events leading to a blackout before it happens. For the latter, there are too many possibilities to analyse power system states, initial events and failures. Therefore, security analyses that try to capture cascading events causing blackouts are often made with probabilistic methods. Most methods use static models, such as cascading overloads, protection failures and voltage collapse. A common simulation model includes static line overloads or voltage instability after contingencies. Although generator protection, controls and dynamic stability often play important roles in blackout evolution, dynamic analysis is seldom applied for reasons of modelling difficulties and complexities. Nevertheless, issues, such as uncontrollable system splitting, angle instability and generation tripping, require dynamic models and simulations [8].

Billinton and Allan made plenty of pioneering research on different aspects of power system reliability, where the focus was on steady-state issues [6, 9]. They pointed out the importance of probability-based reliability evaluation instead of commonly used deterministic criteria that do not take into account the stochastic nature of faults and system behaviour. They also defined concepts, such as hierarchical levels of power system (adequacy) evaluation. The levels are generation alone (hierarchical level I), generation and transmission together (hierarchical level II), and all the parts: generation, transmission and distribution (hierarchical level III) [6, 10, 11].

Billinton and Khan [12] calculated probability indices (the probabilities and frequencies of power system operational states) for composite power systems using a flow chart in detecting the operational states. The states are a normal (secure) state, an alert state and an emergency state. In the normal (secure) state, a power system can withstand single contingencies. In an alert state, the loss of a component will result in a current or voltage violation. The alert state is similar to the normal state in that all constraints are satisfied, but there are no longer sufficient margins to withstand an outage due to a disturbance. In an emergency state, no load is curtailed, but operating constraints have been violated.

The concept of power system states in reliability evaluation is widely used but the definitions of states and sometimes the states are different to some extent. The concept of states represents a classification of complicated reality and this, naturally, can be made in different ways. An example of states different from Billinton's is in Nordic Grid Code [13, p. 66]. In this collection of rules, the power system planning principles were deterministic but they were shifted in a probabilistic direction: more severe consequences are accepted after rare contingencies.

2.3.3 Uncertainty and Dynamics in Power System Analysis

The probabilistic assessments of transient stability were studied for example by Billinton and Kuruganty [14–17], Anderson and Bose [18] and Anders [19]. They considered the randomness of the events that may affect transient instability. These events are for example initial operation conditions, fault type and location, the inertia of synchronous generators, fault duration and critical clearing time. Treating the events probabilistically enables calculating the probabilistic distribution of transient instability.

There have also been security analysis studies made without considering the protection or other substation components. For example, Khan [20], Rei et al. [21] and Leite da Silva et al. [22] carried out analyses of power system security having a probability-based approach. The trips after disturbances occur; the interest is the power system state after those trips. This approach inherently assumes that the protection and circuit breakers act 100% reliably. This assumption is good for operation planning for example, but has limited use as part of an overall approach to the reliability analysis of transmission grids.

Some other approaches exist, too. Miki et al. [23] developed a hybrid model that includes power system dynamic simulations and event trees for protection system operation. The protection systems, but not circuit breakers, are included in the model because "the protection systems play an important role for preventing fault cascading". The protection system is modelled with a Markov model, and the method is applied to a small model grid (19 nodes, 11 lines and 5 generators).

2.3.4 Probabilistic Transitions into Combined Contingencies

Phadke et al. [24] introduced the concept of a *hidden failure* of protection as an important factor leading to cascading outages. A hidden failure is an unwanted and unselective trip by the protection system after the occurrence of another switching event. Since hidden failures are not spontaneous but occur after a switching event, they change an $N - 1$ disturbance into an $N - k$ event and, therefore, are connected to blackout analyses. The hidden failures can influence relay performance in limited areas, called the regions of vulnerability. If an abnormal power system state occurs inside a given region of vulnerability, a hidden failure could cause the relay to incorrectly trip its associated circuit breaker. In order to quantify the effect of hidden failures within a region of vulnerability, a vulnerability index is calculated. The hypothesis is that a probability model for hidden failures enables calculating the events leading to cascading. If given line L and other lines are connected to a bus and any of the other lines trips, a hidden failure in line L might appear: there is probability p that line L will trip (Phadke et al., p. 32).

Several studies have analysed the chains of events that lead to cascading outages and introduced additional concepts, such as critical loading, as a factor that has a remarkable impact on cascading failures. Some examples of the analyses are briefly presented.

Nedic et al. [25] presented a model to analyse the risk of a blackout. Their model estimates the risk of a blackout from a global perspective as a function of critical loading. The blackout is caused by cascading outages due to an initial failure connected with additional trips due to protection malfunctions (sympathetic trips) or generator instability. Sympathetic trips and generator instability were more probable to occur near the original fault. Generator instability was simulated with a heuristic model [26]. They use the concept of critical loading at which the blackout risk increases sharply. Their model, tested for a grid with 1000 buses and 1800 lines, found a critical loading, at which the energy not supplied and the blackout size increased drastically.

Dobson [27], too, has studied the system load increase in the transition from isolated failures to a system-wide collapse. At the critical system loading, the risk of cascading failures starts to increase. With a low system load, the failures can be assumed independent and the blackout probability and size are small. Whilst increasing the system loading, the possible blackout size increases and the probability of cascading outages increases. Dobson also discusses which aspects of cascading failures should be modelled, the tradeoffs between model detail and simulation speed, and what details are required. Possible influencing factors are operational policies, software and human errors. He concludes that there is a need for simple, high-level models to explain the phenomena observed in the simulations.

Kirschen and Jayaweera [28] compared risk-based and deterministic security assessments. According to them, risk-based methods bring considerably more information on which to base operational decisions. In their model, the effect of weather condition is included since, in the case of bad weather, outages occur more often. They estimate the outage costs for customers via Monte Carlo simulations. They apply different operation states, contingencies and possible unwanted unselective trips by relays (sympathetic trips) randomly. They use power flow simulation to determine the system state. If some lines or transformers are heavily overloaded, they are disconnected and the process is repeated, potentially leading to cascading outages. A non-convergence of the power flow computation then indicates voltage instability. In this way, they can evaluate the risk level variations in the same operating conditions during different weather conditions: fair, average and bad weather.

In the Nordic countries, there has been research on the vulnerability of the power system. The scope of one vulnerability analysis [29, 30] was to identify incidents, situations, and scenarios leading to critical or serious consequences to the power system and society as a whole. This study presented a comprehensive methodology for a systematic classification of the Nordic power system with respect to three aspects of vulnerability: energy shortage, capacity shortage and power system failures. The methodology made it possible to quantify the vulnerability of the power system in terms of risk exposures, given by the expected frequency of occurrence of events and their consequences.

2.3.5 Dynamic Event Cascading

Chen [31] and Chen and McCalley [32] estimated the contingency probabilities of power systems by fitting an existing probability model for the historical statistical data. They extend the traditional $N - 1$ contingency to $N - k$ contingencies, where $k > 1$. They developed dynamic event trees for different contingencies. Dynamic event trees differ from normal event trees in three aspects:

1. they include decision nodes where it is possible to take actions for that avoiding or mitigating consequences,
2. they are automatically generated by running the dynamic model and grow according to a set of branching rules and
3. the tree structure, branch probabilities, consequence values, and decisions are updated to reflect changes in the physical network. This means that the evolution of electrical variables defines the possible branch points of the event tree, for example, when a setting value of protection is exceeded.

The dynamic event trees are meant for a decision support tool for control room operators for keeping the system secure. The contingencies and the actions after them change with the topology of the system. Therefore, real-time grid configuration information is required. This is usually possible with energy management systems and state estimation. Possible operator actions are for example generation dispatch, load shedding and system islanding.

Levi et al. [33] did a reliability analysis, which included dynamics. They calculated post-fault states with a Monte Carlo method and doing so, added the calculation of the states with dynamic simulations into the reliability model of Khan and Billinton [34]. The steps in the simulation were follows:

1. transient stability analysis,
2. the evaluation of frequency dynamics during governor control,
3. activation of the emergency level of the thermal protection and the steady-state after the automatic generation control (AGC),
4. the minimization of the overall curtailed load,
5. overload rotation and
6. voltage restoration and economic dispatching.

They analysed the 39-node IEEE test system [35], and their results indicate that the conventional steady-state reliability analysis gives too optimistic results, since it ignores a number of unfavourable power system phenomena.

2.3.6 Conclusion

There exists no standard methodology for power system *security* analysis. There are different methods adapted to specific system characteristics but none of which can capture the system security problem as a whole. System dynamics is seldom included; usually a static post-fault steady-state analysis is selected. Developed methods rely on simplifications in modelling and definite scope is necessary.

For adequacy, several measures like the loss of load probability (LOLP), system average interruption duration index (SAIDI), system average interruption frequency index (SAIFI), etc. are available for distribution grids.

The power system reliability analyses are often made without a substation model or with a limited model. Sometimes the methods use probability values for switching actions, which occur at the substation and change the grid topology. Protection rather than circuit breaker actions are included.

For large complex systems, it may not even be worthwhile to try to include all aspects in one methodology.

2.4 Security and Electricity Markets

As already stated, large disturbances, e.g. partial or total system breakdowns, have effects on the critical functions of modern society. These events should be avoided and the desired level of security maintained in liberalized electricity markets, too.

Nowadays it seems to be a common thought that because of electricity markets, blackouts occur more often. The reason behind this thought may be that electricity markets lead to a more efficient use of the existing networks and increased mutual interdependency. The markets alone should not be blamed, but rather the grid operation not adapted for the new procedures. In Italy and the United States in 2003, the power flows were high before the blackouts. The overloading of remaining lines after several trips caused instability and the collapse. In these cases, high power flows were amongst the causes of the blackout. A report by the IEEE PES CAMS Task Force [8] reveals that blackouts do not always occur during peak load days in winter and summer, when the system is stressed but rather on "shoulder" periods (spring and autumn), and are caused by outages due to maintenance or other reasons. The combination of these outages changes the power flows and dynamic characteristics of the system. The result may be a much higher probability of a cascading outage due to the unexpected forced outage of other equipment or to operating mistakes.

Roles and responsibilities affecting security are defined in the legal framework. In electricity markets, the responsibility to ensure secure power system operation falls to those parties who develop, maintain and operate the power system. Usually, transmission system operators are responsible for the security of the power system as a whole within their region—typically a country or part of a country. Distribution system operators are responsible for ensuring security in their networks and for meeting the requirements set by the transmission system operator for security in order to avoid local faults spreading in other parts of the national power system.

The system operators also have powers to set requirements for the connection and operation to the parties connected to their grid—generators and loads—to ensure a secure operation of the power system. The legal framework may set performance requirements for system operators through obligations and/or incentives to maintain a predefined level of security and quality of electric supply. The system operator copes with these legal obligations by developing, maintaining and operating the grid.

Transmission system operators apply congestion management methods of limiting the power flows in the grid to ensure secure system operation. In Europe, the legal framework requires that these methods should be market-based and the transmission capacity should be given to those who value the capacity most. The core of congestion management methods is the calculation of transmission capacity taking into account probable power flows, faults and consequences of these faults to the system. The legal framework may require that the transmission capacity is firm when it is given to the market. Specific legal instructions may be applied to allow curtailments of the transmission capacity and/or reduced firmness of the transmission capacity.

The transmission capacity made available to the electricity market should be maximized without violating the system security. This requirement for a more efficient use of the grid calls for advanced methods and approaches to evaluate the security of the power system. Here the PSA approach may give added value to those planning, maintaining and operating the power system.

Integration of electricity markets implies that more transmission capacity will become available between national markets. This requires that either more interconnections should be built between national power systems or more transmission capacity should be released to the electricity market using the existing interconnections. The national power systems become thus more dependent on each other due to the tighter electric connection and due to the increased flows between power systems. Furthermore, the variable generation to meet the policy goals for sustainability will require more flexible power systems. It will increase the interdependence between systems due to the larger variations in the power interdependency of national power systems. This implies that a disturbance within a national power system may extend to neighbouring power systems and in the worst case cause a system breakdown extending several national power systems [36–38].

2.5 Scope of the Book

The PSA approach for transmission grid security presented in this book is one of the security analysis methods for specific purposes; it is a probabilistic approach to assessing the risk of a system breakdown after grid faults, and observing system dynamics, too [39, 40]. Based on systematic and analytical modelling of post-fault events rather than on random sampling, the PSA approach can reveal the vulnerable parts of the system. The main objective is to increase the understanding of the system and reveal those components, where the improvements most effectively support the security after grid faults.

The PSA approach enables the inclusion of dynamics in the analysis. The methods that rely on steady-state analyses after faults cannot capture the system level dynamic stability phenomena though this would be beneficial and necessary when dynamics and stability after contingencies are crucial. This is typically the case when power flows are high and transmission routes are long. In this case, the *transmitted* rather than *consumed* power is important for the system security.

The power system reliability analysis is a large and complex issue. There are different methods, none of which can capture the whole problem. Commonly used methods for steady-state adequacy analyses exist, but security methods are adjusted for specific properties of the system under study.

A common measure and the important matter for transmission system security is the system state (such as secure, alert, collapsed), rather than the states of the connection points of single customers. With the PSA method, it is possible to analyse different chains of events in such a way that their consequences are quantitatively comparable and their effect on the system state can be found. Central to the approach presented in this book is the system security including an analytical probability model for post-fault substation events (the protection system and circuit breaker operations), and the power system dynamics after the contingencies.

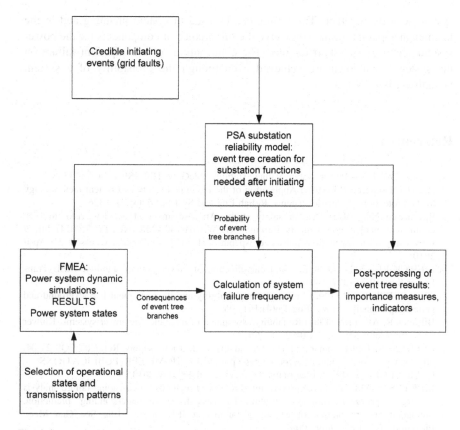

Fig. 2.4 A simplified block diagram of the PSA security analysis of transmission grids [40]

The method described in this book and presented in Fig. 2.4 is applicable to real-size transmission grids. The model for substation post-fault operations uses event and fault trees and, therefore, inherently introduces the possibility of calculating different grid level importance measures for substation components and for model parameters. With the component and parameter importance measures, the relatively effective ways of improving grid security can be found. They also help to find the contributing factors to a system breakdown. The method takes into account the effect of the following issues on reliability:

- the frequency of faults,
- fault locations on the line,
- different substation arrangements,
- the failure rates of substation components and
- the dynamic behaviour of the power system after different contingencies.

Resources (always limited) can be used in a more efficient way after the contributions of components to a possible blackout are known. Therefore, the method can be the basis for connecting reliability-centred maintenance and power system security together. The results can be used in system planning and in the maintenance planning since they give the importance of components for the power system security after short circuits. The contributions of a component failure on the system breakdown are achieved even though the probability of a system breakdown is very low.

References

1. Taylor CW (1994) Power system voltage stability. McGraw-Hill. ISBN 0-07-113708-4
2. Hortal J, Izquierdo JM (1996) Application of the integrated safety assessment methodology to the protection of electric systems. Reliab Eng and Syst Saf 52(3):315–326
3. Rasmussen NC (1975) Reactor safety study. An assessment of accident risks in U.S. commercial nuclear power plants. Rep numbers WASH-1400-MR (NUREG-75/014). http://www.osti.gov/energycitations/purl.cover.jsp?purl=/7134131-wKhXcG/. Accessed 21 April 2010
4. IEEE/CIGRE (2004) Definition and classification of power system stability. IEEE Trans Power Syst 19(3):1387–1401
5. Allan R, Billinton R (1992) Power system reliability and its assessment I. Background and generating capacity. Power Eng J 6(4):191–196
6. Billinton R, Allan RN (1988) Reliability assessment of large electric power systems. Kluwer Academic Publishers. ISBN 0-89838-266-1
7. CEER (2008) 4th Benchmarking report on quality of electricity supply. Ref: C08-EQS-24-04. http://www.energy-regulators.eu/portal/page/portal/EER_HOME/EER_PUBLICATIONS/CEER_ERGEG_PAPERS/Electricity/2008. Accessed 13 May 2010
8. IEEE PES CAMS Task Force (2008) Initial review of methods for cascading failure analysis in electric power transmission systems. In: Task force on understanding, prediction, mitigation and restoration of cascading failures at IEEE Power Eng Soc Gen Meet, Pittsburgh, PA, USA, July 2008
9. Billinton R, Ringlee RJ, Wood AJ (1973) Power system reliability calculations. MIT Press. ISBN 0-262-02098-X
10. Billinton R, Allan RN (1984) Power system reliability in perspective. Electron Power 30(3):231–236
11. Allan R, Billinton R (2000) Probabilistic assessment of power systems. Proc IEEE 88(2):140–162
12. Billinton R, Khan E (1992) A security based approach to composite power system reliability evaluation. IEEE Trans Power Syst 7(1):65–72
13. Nordel (2007) Nordic Grid Code 2007. http://www.entsoe.eu/index.php?id=62. Accessed 13 May 2010
14. Billinton R, Kuruganty PRS (1979) Probabilistic evaluation of transient stability in a multimachine power system. Proc Inst Electr Eng 126(4):321–326
15. Billinton R, Kuruganty PRS (1980) A probabilistic index for transient stability. IEEE Trans Power Appar Syst PAS-99(1):195–206
16. Billinton R, Kuruganty PRS (1981) Probabilistic assessment of transient stability in a practical multimachine system. IEEE Trans Power Appar Syst PAS-100(7):3634–3641
17. Kuruganty PRS, Billinton R (1979) An approximate method for probabilistic assessment of transient stability. IEEE Trans Reliab R 28(3):255–258
18. Anderson PM, Bose A (1983) A probabilistic approach to power system stability analysis. IEEE Trans Power Appar Syst PAS-102(8):2430–2439
19. Anders GJ (1990) Probability concepts in electric power systems. Wiley. ISBN 0-471-50229-4

20. Khan ME (1998) Bulk Load Points Reliability Evaluation Using a Security Based Model. IEEE Trans on Power Syst 13(2):456–463
21. Rei AM, da Silva AML, Jardim JL, de Oliveira Mello JC (2000) Static and Dynamic Aspects in Bulk Power System Reliability Evaluations. IEEE Trans on Power Syst 15(1):189–195
22. Leite da Silva AM, Endrenyi J, Wang L (1993) Integrated Treatment of Adequancy and Security in Bulk Power System Reliability Evaluation. IEEE Trans on Applied Superconduct 3(1):274–285
23. Miki T, Okitsu D, Kushida M, Ogino T (1999) Development of a Hybrid Type Assessment Method for Power System Dynamic Reliability. IEEE Int Conf on Syst. In Proceedings: Man and Cybern 1999, SMC '99 Conf Proc 1:968–973
24. Phadke AG, Horowitz SH, Thorp JS (1995) Anatomy of Power System blackouts and Preventive Strategies by rational supervision and control of protection systems. ORNL/Sub/ 89-SD630C/1, A rep for the Power Syst Technol Program, Energy Division Oak Ridge Natl Lab
25. Nedic D, Dobson I, Kirschen D, Carreras B, Lynch V (2006) Criticality in a cascading failure blackout model. Electric Power and Energy Syst 28:627–633
26. Rios MA, Kirschen DS, Jayaweera D, Nedic DP, Allan RN (2002) Value of Security: Modeling Time-Dependent Phenomena and Weather Conditions. IEEE Trans on Power Syst 17(3):543–548
27. Dobson (2007) Where is the edge for cascading failure? Challenges and opportunities for quantifying blackout risk. In: IEEE Power Eng Soc Gen Meet, 24–28 June 2007, USA
28. Kirschen DS, Jayaweera D (2007) Comparison of risk-based and deterministic security assessments. IET Gener Transm Distrib 1(4):527–533
29. Doorman GL, Kjølle GH, Uhlen K et al (2004) Vulnerability of the Nordic power system. Report to the Nordic Counc of Ministers, TR A5962. http://193.88.185.141/Graphics/ Energiforsyning/Forsyningssikkerhed/Elforsyningssikkerhed/ VulnarabilityoftheNordicPowerSystem.pdf. Accessed 13 May 2010
30. Doorman GL, Uhlen K, Kjølle et al (2006) Vulnerability analysis of the Nordic power system. IEEE Trans Power Syst 21(1):401–410
31. Chen Q (2004) The probability, identification, and prevention of rare events in power systems. PhD dissertation, Iowa State University, Ames, Iowa. http://www.pserc.org/ cgi-pserc/getbig/publicatio/2004public/qimingchen_phd_dissertation_on_cascading.pdf. Accessed 13 May 2010
32. Chen Q, McCalley JD (2005) Identifying high risk N-k contingencies for online security assessment. IEEE Trans Power Syst 20(2):823–834
33. Levi VA, Nahman JM, Nedic DP (2001) Security modeling for power system reliability evaluation. IEEE Trans Power Syst 16(1):29–37
34. Khan ME, Billinton R (1992) A hybrid model for quantifying different operating states of composite power systems. IEEE Trans Power Syst 7(1):187–193
35. Grigg C, Wong P, Albrecht P et al (1999) The IEEE reliability test system—1996. IEEE Trans Power Syst 14(3):1010–1020
36. UCTE (2007) Final report—system disturbance on 4 November 2006. http://www.entsoe.eu/ index.php?id=59. Accessed 13 May 2010
37. UCTE (2004) Final Report of the investigation committee on the 28 September 2003 Blackout in Italy. http://www.entsoe.eu/index.php?id=59. Accessed 13 May 2010
38. U.S.–Canada Power System Outage Task Force (2004) Final report on the August 14th Blackout in the United States and Canada: causes and recommendations. https://reports.energy.gov/. Accessed 13 May 2010
39. Pottonen L (2005) A method for the probabilistic security analysis of transmission grids. A doctoral dissertation, Helsinki University of Technology, 951-22-7591-0, 951-22-7592-9 http://lib.tkk.fi/Diss/2005/isbn9512275929/. Accessed 29 June 2010
40. Haarla L, Pulkkinen U, Koskinen M, Jyrinsalo J (2008) A method for analysing the reliability of a transmission grid. Reliab Eng Syst Saf 93(2):277–287

Chapter 3
Basic Concepts of Transmission Grid Planning

3.1 Introduction

The Chapter describes transmission planning from the long-term time frame until the real-time system operation, presents factors that set the limits to the transmission capacity, and presents the most important components of the transmission system.

Electric power systems play an important role in a modern society supplying electricity continuously to diverse applications. Conventionally, the power system has been functionally divided into three parts, the generation system, the transmission system and the distribution system. The generation system converts other forms of energy, for example, the potential energy of water or energy of the fuels, with the help of prime movers and generators into electricity. The transmission systems with high voltage and large capacity lines, most often with overhead lines and underground cables, take care of the bulk power transmission of electricity over long distances from generation sites to consumption areas. Distribution systems finally deliver electricity to individual consumers with low voltage lines. Electric power systems cover virtually all populated areas, and the majority of peoples are dependent on services electricity supply facilitates.

What is characteristic of alternating current power systems is that a balance must be kept between electricity generation and consumption at every moment since in practice large-scale electricity storages do not yet exist. This sets requirements for the reserve capability of generation apparatus to follow the inherent variation patterns of electricity consumption.

Reliable operation, control and planning of such an extensive and dynamic system are real challenges. Section 3.2 presents the purpose and practices of transmission system planning. Section 3.3 describes the phenomena and criteria that set the limits to the transmission capacity in the meshed grids. After that, Sect. 3.4 presents reliability as a function of power. Finally, Sect. 3.5 describes the important components of the grid for the reliability analysis, like lines and substations.

L. Haarla et al., *Transmission Grid Security*, Power Systems,
DOI: 10.1007/978-0-85729-145-5_3, © Springer-Verlag London Limited 2011

3.2 The Purpose of Transmission System Planning

3.2.1 Scope

The objective of planning is to provide sufficient transmission capacity in the long run efficiently implemented and on right time observing safety, environmental, system security and socio-economic aspects as well. In order to be able to find a balance between these partly contradictory requirements, it is necessary to put a price on all these factors.

Efficiency comes from the investments that provide the biggest increase on transmission capacity per money spent. Observing safety requirements is a must originating from relevant legislation. Reliability can be quantified by impacts on customers if the service is not available but usually in transmission system planning, also a deterministic security rule is applied. Nowadays, environmental aspects are often evaluated in an environmental impact assessment procedure or similar, required by the legislation, although the quantification of environmental factors is usually difficult.

The system reinforcement will change electricity market conditions as well by increasing transmission capacity. The impact of the capacity increase for producers and consumers can be estimated with multi-area market model simulations comparing the case with the planned reinforcement to the case without reinforcement. All in all, the following socio-economic consequences following the capacity increase may be considered.

1. changes in producer and consumer benefits and bottleneck revenues,
2. changes in risk for energy rationing,
3. changes in risk for short of power situations,
4. changes in active power losses,
5. the reduction of possibilities to misuse market power and
6. changes in ancillary service trading.

Many factors can endanger reliability of the electricity supply service to customers. Equivalently with the principal scheme of the entire power system, also reliability threats can be divided into three categories:

1. problems in generation,
2. failure in transmission or
3. a fault in distribution or in local connection.

The reason for a supply interruption to a customer may reside in any of these parts although typically the distribution system is responsible for the majority of unavailability. Problems in generation and transmission systems on the other hand can affect on a larger number of customers. Either way the harm for a customer can be estimated with power interrupted, the duration of the interruption and with the customer specific value of electricity not supplied. Generally, the value of energy not supplied per energy unit will decrease when the electricity intensiveness of the customer production process increases but will remain higher than the selling price of electricity.

3.2.2 Generation and Grid Adequacy

In the reliability analysis of the composite electric power system, the requirement of sufficient overall generation capacity to cover the consumption at every moment can be seen as the first step. Obviously, the most critical situations are periods of high consumption of electricity. In generation capacity planning, a measure for the adequacy of generation is the loss of load probability (LOLP). [1] explained the procedures of calculation the LOLP and gave examples.

The focus of this book is the analysis of the transmission system and its role in composite power system reliability. If the generation fails to meet the demand for electricity, then transmission and distribution systems cannot help the situation unless they provide access to external generation sources. On the other hand, if the transmission capacity from generation resources to distribution and consumption areas is inadequate, then this cannot be solved either with activation of extra generation capacity unless that capacity happens to locate geographically in the deficit area. Actually, this is the main task of transmission system planning, to connect multiple power generation sites with the multitude number of bulk consumers at remote locations in a cost-effective way with adequate transmission capacity observing safety, reliability and environmental considerations. Compared with generation capacity planning, a new factor in transmission system planning is the distance between generation and consumption and the fact that transmission paths do have restricted capacities. This important distinction is illustrated in Fig. 3.1.

The salient idea having the facilitated development of the liberalised electricity market is that market access is based on the point tariff eliminating the impact of the transmission distance on market actors. Physically, the situation is naturally unchanged, still adequate transmission capacity from generators to consumers and between different market areas should be provided. This obligation is appointed to transmission system operators, whose duty is to connect clients, ensure secure system operation and to develop the transmission system according to clients' reasonable needs.

Whilst analysing the reliability of the main transmission grid of the electric power system, it is obvious from earlier presented that neither customer nor generation oriented indices are optimal for that purpose. Unlike the overall purpose of the power system to serve the customers, a primary objective of the main transmission system can be seen as to keep the system intact even by sacrificing a part of generation or loads. One measure of failing in this task is the number of those incidents where the integrity of the system is lost in an unplanned manner. That kind of an incident could be either a total blackout of the system or just an

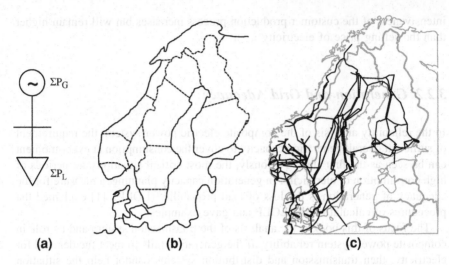

Fig. 3.1 Different perspectives of a power system planning **a** presents pure generation capacity planning with a single busbar model **b** equals multi-area generation capacity planning recognising restricted transmission capacity between separate generation areas and **c** presents true transmission system planning observing real transmission lines and substations connecting generation and consumption

extra disconnection of load or generation in addition to planned disconnection in a contingency.

3.2.3 Adequacy and Security

It is important to distinguish the difference between two commonly used reliability terms—adequacy and security—in context of electric power systems.

In transmission systems, *adequacy* means the ability of the system to meet all credible power flow situations constantly without overloading and voltage violations. In a transmission system, there are adequate assets in place to cope with all assumed power flows [2].

A transmission system is *secure* if it in addition also tolerates all the $N - 1$ contingencies without losing its stability and maintaining voltages and system frequency within the predefined limits of post-fault situations [2].

3.3 Transmission Capacity

At first, the meaning of transmission capacity in the main transmission systems is illustrated. Factors affecting and limiting transmission capacity in electric power grids are then shortly described.

3.3.1 N − 1 Criterion

In transmission system planning, a common criterion is the so called $N - 1$ criterion. Shortly, this means that a system having N components can continue its operation even though whichever single component at a time is taken out of operation. However, there is variation how the actual process leading to an outage of a single component can develop and which components are observed. These differences can be mainly explained with different system characteristics and environment conditions. The selection of contingencies for the $N - 1$ criterion is usually based on the probabilistic evaluation of the fault frequencies of the diverse power system components and possible consequences of those faults. In general, larger consequences can be allowed for rare contingencies, but the most common contingencies should be imperceptible for grid clients.

In practice, the criterion means that a trip of a component after a fault does not affect the consumers and the grid remains stable and there is no overloading in the transmission system. However, from the operational point of view, the security of the system is decreased for a moment after a fault, and it may be necessary to adjust power transfers and activate reserves to cope with the next possible fault. This kind of operational situation is called an alert state.

An implication of the $N - 1$ criterion is that transmission capacity of the system is not the sum of capacities of its parallel components, but a sum capacity of parallel components decreased with the capacity of the strongest one component. A special case is the first step towards a meshed system when one decides to close a loop i.e. extend a radial transmission path with an alternative supply direction. The second line connection does not increase transmission capacity when $N - 1$ criterion is introduced; it merely improves the dependability of the supply path. The next parallel transmission path extensions are then rewarded also with increased transmission capacity since the $N - 1$ criterion is already fulfilled.

3.3.2 Factors Limiting Transmission Capacity

In electric power systems, stability is interpreted as the ability of the system to maintain steady operation without power fluctuations and angle swinging. An essential part of steady operation is also that the system is capable of maintaining adequate voltage levels and balances between generation and consumption. The requirement for stability applies both in continuous operation (a sort of quasi-stationary state) and during contingencies. The joint working group of organisations IEEE and CIGRE [2] divided the stability in different categories shown in Fig. 3.2. More about stability can be found e.g. in the books of [3] and [4].

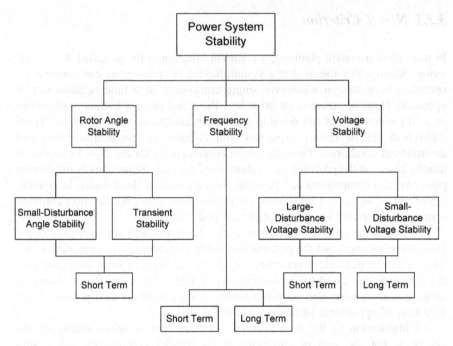

Fig. 3.2 Classification of power system stability by the IEEE/CIGRE joint task force [2]

3.3.2.1 Rotor Angle Stability

Rotor angle stability can be illustrated and explained with *power angle equation* and *equal area criterion*.

The power angle equation

$$P = \frac{EU}{\Sigma X} \sin \delta \qquad (3.1)$$

gives the transmitted electrical power P as a function of voltages E and U, the series reactance of the transmission path ΣX and angle difference δ. Here, the angle difference is between the rotor angle of a single generator and the 'rotor' angle of the equivalent generator representing a large system.

The mechanical power of the turbine-generator set equals electrical power transmitted to the power system in a steady state operation with initial rotor angle δ_1 in Fig. 3.3. During a fault, for example, a line short circuit, the voltages are low and, therefore, also the electrical power is low according to Eq. (3.1) causing an unbalance between the mechanical power and the electrical power. Therefore, the shaft of the turbine-generator set accelerates. The generator angle starts to increase rapidly. The line protection disconnects the faulty line with the rotor angle δ_2 in Fig. 3.3. Now electrical power starts to follow the post-fault power curve eventually reaching values higher than the mechanical power and the turbine-generator

Fig. 3.3 Equal area criterion for transient stability when a single generator is connected to a large system d is the angle difference between the voltages E and U **a** shows the connection and **b** presents the power angle curves before, during and after the short circuit

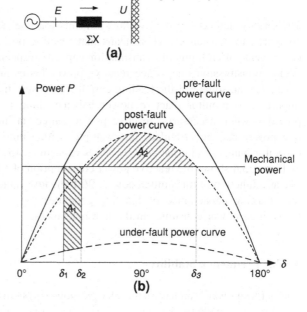

set starts to decelerate. Stability will be maintained if the accelerating energy does not exceed the decelerating energy. In such a case, the rotor angle of the generator will reach its final value δ_f through oscillations. Oscillations are inherent to synchronous machines because of inertia of the rotating turbine-generator sets.

The equal area criterion, presented in Fig. 3.3, defines a transient stability criterion for the single machine case with the help of power angle equations for pre- and post-fault conditions and the equation during the fault, too. The system can maintain transient stability if the accelerating area A_1 is less or equals the retarding area A_2, i.e. if the angle did not grow too much during the fault and if the electrical power after the fault is sufficient. How large a value the rotor angle can reach during the fault and how large the accelerating energy is, depends on the fault clearance time. Faster protection decreases the clearance time. For the post-fault situation crucial is, how much the transmission path has weakened i.e. the resulting reactance has increased. If the post-fault electric power curve cannot reach high enough values compared with the mechanical power, the retarding area cannot compensate the accelerating area and stability will be lost.

For multi-machine systems and for longer time periods than just the first swing, more detailed modelling is required for assessing stability. Nevertheless, these basic principles help to understand the essential factors for maintaining angle stability. In practice, stability is assessed with power system simulations with appropriate software utilizing numerical step-by-step methods for the linearization of highly non-linear behaviour of the actual systems with a large number of state variables.

3.3.2.2 Voltage Stability

The series reactances of high voltage transmission lines and transformers are high compared to resistances. This means that reactive power consumption of these components often limits the transmission capacity, especially with long and heavy loaded transmission lines since reactive power losses are in relation to the line reactance and square of the loading current. To ensure transmission capacity, it is important to monitor reactive power balance along the transmission path and provide—when necessary—reactive power sources at intermediate points. Reactive power can be provided, for example, with shunt devices like generators, capacitor banks or alternatively by series capacitors. [5, 6]

Figure 3.4 shows the reactive power consumption of a typical 400 kV overhead line as a function of transmitted power. With the low values of power transmission, the charging capacitance of the line generates more reactive power than the transmitted power consumes in the line reactances.

3.3.2.3 Frequency Stability

The active power balance of the electric power system has an influence on the frequency of the alternating current. The deficit of power will decrease the rotation speed of the synchronous machines of the system and surplus of power will increase it. Synchronous generators are equipped with speed governor control to regulate the power of their prime mover to keep the system frequency at its nominal value within their free capacity and predefined droop settings. Crucial for maintaining adequate frequency for the synchronous operation of the power system is how much generation can be lost either through a generator tripping or as a consequence of a single fault in transmission system disconnecting surplus area from the rest of the power system. Equal amount of primary reserves must be

Fig. 3.4 The consumption of reactive power (Q) of a 100-km long 400 kV overhead line as a function of power transmission (P)

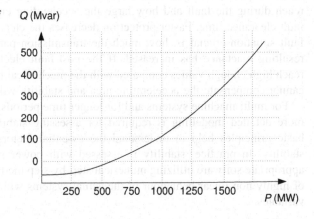

available at every moment in operation to avoid too large frequency decline causing system separation and island operation.

3.3.2.4 Thermal Capacity

The thermal capacities of cables, overhead lines and substation equipment can be a limiting factor in strongly meshed networks with the relative short transmission distances.

Usually conventional alternating current components like overhead lines and power transformers can be temporary loaded over continuous rating values giving margins for short time operation during contingencies and exceptional situations. For those situations, it is, however, important to ensure beforehand sufficient resources for system operators to restore power flows within the given time frame.

3.4 Reliability Evaluation in Long-Term System Planning and in System Operation

3.4.1 Challenges

It is important to recognise different viewpoints and objectives of reliability evaluation of transmission systems in its separate stages of life cycle from long-term strategic planning to the real-time operation of the system.

The actual reliability of a transmission system will realise in operation of the system. In practice, at the operation stage, the reliability of the transmission system can be increased mainly by reducing power transfers. This is not favourable due to adverse effects on the functioning of the electricity market or restrictions to the power input or output of the customers.

Fundamental decisions concerning the system reliability have already been made in the process of system planning. Examples of such long-term system decision are

- to what extent redundancy should be built in the transmission system and
- for how large capacities transmission paths between different sub-systems will be constructed.

In the long-term system planning, the uncertainty of the future development and demand for transmission capacity complicates the development task of a transmission system. Usually, this is tried to overcome by applying scenario techniques to probe the required expansion projects in a transmission system under different possible development paths. The transmission system reinforcements that are common for multiple scenarios are then the best bets for closer project planning.

3.4.2 *Reliability as a Function of Power Flow*

Let us consider two power systems, which have interconnections consisting of several parallel lines with intermediate substations shown in Fig. 3.5a. In a situation where the power exchange between the systems is minimal or near zero, both systems are self-sufficient and independent of the reliability of the transmission path. When power transmission is increased, the receiving system becomes gradually more and more dependent of the transmitted power and the transmission path.

One threshold is when the amount of imports over the transmission path exceeds the local reserves. If the import of electricity interrupts for some reason, then the receiving system will no more regain balance with its local generation resources but at least part of the consumption will experience an interruption when load shedding takes place. Even a blackout may occur in the receiving system if no defence schemes exist.

Equally, with generation reserves, also the availability of transmission reserves can be seen as a turning point since electricity imports from the neighbouring system is an alternative to the local generation resources. In such a situation, it is apparent to set a requirement for the reliability of the transmission path since the receiving system is dependent on it.

Hence, the reliability of the transmission path is a function of power flow on that path and the reliability will decrease when the power flow is gradually increased. In Fig. 3.5, an example of reliability analysis of a transmission path is shown. The curve can be considered as an indicator of the momentary risk of the breakdown of the power transmission path and the receiving subsystem. The risk increases stepwise when single contingencies are able to collapse the power system. The risk increases further on when lighter and more probable contingencies become critical for the operation of the transmission path.

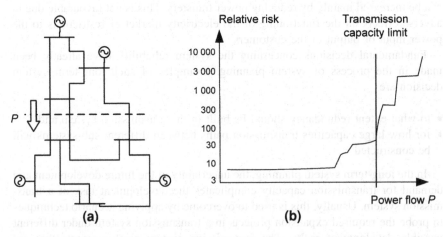

(a)   **(b)**

Fig. 3.5 Momentary major disturbance frequency of the receiving power system as a function of power flow on a transmission path between two systems

The most severe $N - 1$ fault i.e. the single fault, which requires the lowest value for transmission in order the system to survive the contingency is called the *dimensioning fault*. Transmission capacity for the system is defined with the dimensioning fault and is illustrated in Fig. 3.5b.

In reality, it is not so easy to identify all factors affecting transmission system reliability and figure out typical interactions between them. The power system operating point is characterised by the amount of generation, load and power flows. Furthermore, the conditions on adjacent cross sections will have impact on the transmission path under study, therefore, the actual megawatt values of knee points will vary somewhat from case to case.

3.4.3 Average Reliability as Result of Momentary Risk and Density Function of Transmission

A logical consequence of the previous assumption, that a momentary risk of major disturbance in a power system is a function of power transmission, is that the average risk over a certain period is then momentary risk multiplied with the probability density function of the transmission on that transmission path. An example of that kind of density function of the transmission is shown in Fig. 3.6. The hourly mean values of the power flows of 1 month are grouped in categories. There are 720 samples in total, and the grand total of the probabilities of all categories is one.

3.5 Components of the Transmission Systems

The main components of the transmission grid are different lines (overhead lines, underground and submarine cables), and substations, where transformers, circuit breakers, protection and control devices are located.

3.5.1 Transmission Lines

3.5.1.1 Overhead Lines

Overhead lines are prone to external hazards and nature phenomena. Although structures are relative simple and dependable, large geographical area and quantity—easily thousands of line kilometres—mean that disturbances every now and then do occur. As an example of external hazards are crashes of vehicles, lightning strokes, forest fires and ice storms. Faults caused by atmospheric over-voltages due to lightning strokes are common on overhead lines. An advantage of

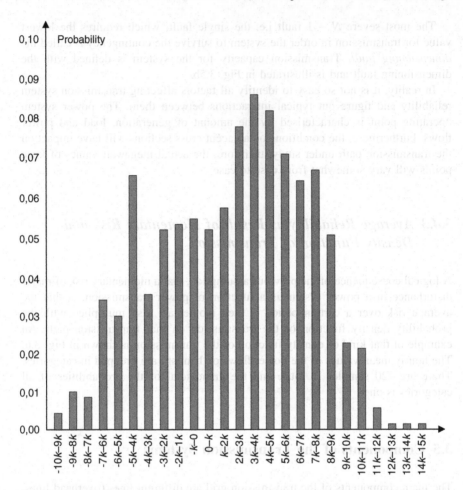

Fig. 3.6 An estimate of the monthly probability density function of the transmission levels, grouped in the multiples of k in the example system of Fig. 3.5a

overhead lines is that the insulating material—air—will recover from a breakdown caused by lightning after the ionisation has vanished from the air gap. Damages that are more serious can follow from crashes with vehicles if a tower collapses. Ice storms may cause the collapse of large line sections but fortunately, severe storms are rare Fig. 3.7.

3.5.1.2 Underground and Submarine Cables

Cables are better shielded against external hazards and do experience less faults per circuit length. Often this comes at the expense of accessibility, when a failure happens the localization of the exact fault spot is usually slower than with overhead lines and dismantling of shielding prolongs reparation time.

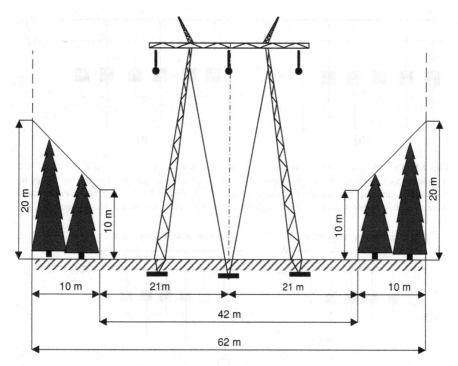

Fig. 3.7 A cross section of 400 kV overhead line right-of-way with guyed single-circuit towers

3.5.2 Substation Schemes and Components

3.5.2.1 Substation Arrangements

As the nodes of the transmission grid, substations allow the implementation of selective transmission grid protection, switches, i.e. circuit breakers, for changing the grid topology and grouping of feeders for desired configurations. The flexibility and features will vary depending on the selected substation scheme. In the following, some typical substation arrangements are presented.

Figure 3.8 presents two different busbar arrangements. A double busbar arrangement with one circuit breaker at each line end (Fig. 3.8a) is less costly. Having two circuit breakers at each feeder (Fig. 3.8b) is more expensive, but it gives flexibility for the system. For example, circuit breaker maintenance is easier, and the arrangement keeps the grid intact after busbar faults. Even though the faulty busbar trips, all lines remain via the healthy busbar.

After a line fault, one or two circuit breaker per line end should trip, depending on the busbar arrangement. If a circuit breaker fails to trip, then the breaker failure relay trips all the other circuit breakers connected to the same busbar as the stuck circuit breaker and, in some cases, it also trips the circuit breakers at the other line

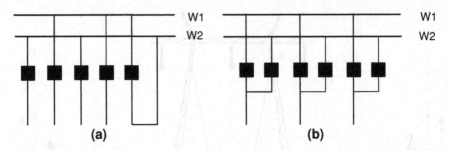

Fig. 3.8 Two different double busbar arrangements with two main busbars W1 and W2 **a** presents a case where, in a normal operation, each line is connected either to busbar W1 or to busbar W2, and the busbars are connected to each other with a bus coupler **b** presents a case where a line is connected to each busbar with a circuit breaker; no bus coupler is needed

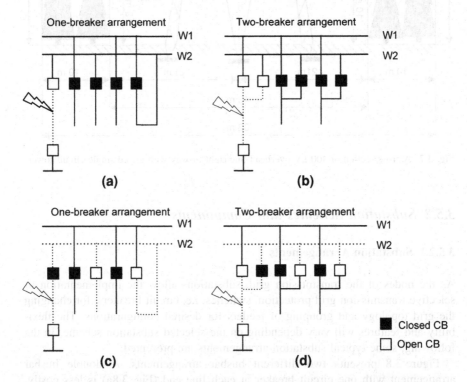

Fig. 3.9 Circuit breaker states after a line fault with two different busbar arrangements **a** and **b** present the correct operation of the circuit breakers of the faulted line **c** and **d** present corresponding situations after the breaker failure protection has isolated the busbar W2 to which the faulted (stuck) circuit breaker was connected. *Dotted lines* represent the dead grid sections. The circuit breaker at the remote line end operates correctly in all cases

end. Figure 3.9 presents how the breaker failure relays isolate the busbars with the corresponding busbar arrangements as in Fig. 3.8 if the line breaker of the faulty line fails to open. In Fig. 3.9c, also the circuit breaker at the other line end trips. A

Fig. 3.10 The transmission grid is divided into protection sections. The protection system can disconnect each section from the grid if a fault occurs on that section. In the figure, the sections are drawn with *dash lines* and the *black squares* denote circuit breakers

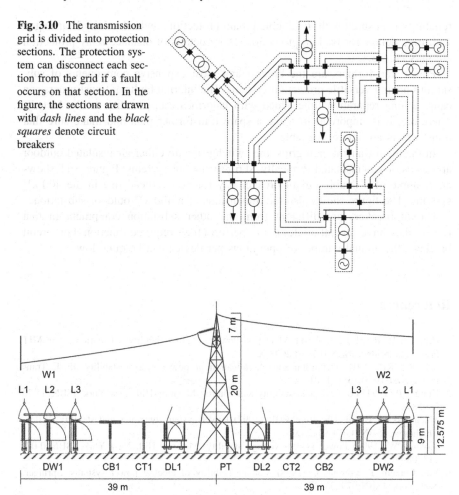

Fig. 3.11 An example of the cross section of 400 kV two-breaker bay on an outdoor switchyard. W1 and W2 are the main busbars. DW1 and DW2 are pantograph disconnectors under the busbars, CB1 and CB2 are circuit breakers, CT1 and CT2 are current transformers, DL1 and DL2 are line side disconnectors and PT is a potential transformer in this overhead-line bay. The total length of the bay is some 80 m

stuck circuit breaker at one-breaker substation causes the loss of lines, but a similar case at a two-breaker substation leaves the lines in use.

3.5.2.2 Substation Equipment

Circuit breakers divide the transmission system into protected sections. For example, lines, transformers and substation busbars are such protected sections. Figure 3.10 illustrates the protected sections of the transmission grid. High reliability is ensured with the doubled main protection, and selectivity is arranged with time delays for backup protection. More details of distance relay protection are in Sect. 6.3.1.

Power transformers are the biggest and most expensive components of sub-stations having versatile protection systems. The internal faults of transformers can cause a long reparation time added with laborious transportation to the factory. Therefore, it is important to make a spare transformer plan and make sure that transformers are interchangeable.

In the main transmission grids, the switchgears are either air insulated outdoor apparatus or gas insulated (typically SF6-gas) indoor systems. Figure 3.11 shows an example of a double circuit breaker bay for an overhead line in the 400 kV systems. The figure shows also the dimensions of a 400 kV outdoor substation.

Circuit breakers are different from the other substation equipment in that respect they have several modes not to perform their required function. For circuit breakers, the annual number of operations per device can be quite low.

References

1. Billinton R, Ringlee RJ, Wood AJ (1973) Power system reliability calculations. The MIT Press, Cambridge, ISBN 0-262-02098-X
2. IEEE/CIGRE (2004) Definition and classification of power system stability. IEEE Trans Power Syst 19(3):1387–1401
3. Kundur P (1994) Power system stability and control. McGraw-Hill, New York, ISBN 0 070 35958 X
4. Machowski J, Bialek JW, Bumby JR (1997) Power system dynamics and stability. Wiley, Chichester, ISBN 0 471 97174X
5. Taylor CW (1994) Power system voltage stability. McGraw-Hill, New York, ISBN 0-07-113708-4
6. Van Cutzem T, Vournas C (1998) Voltage stability of electric power systems. Springer, New York, ISBN 0792381394

Chapter 4
Basic Concepts of Reliability Theory

4.1 Introduction

In the previous chapters, the issue of transmission grid security was presented. The insight likely to be gained from a probabilistic look at the risk analysis problem of the transmission power system was explained. Still, connecting combinations of occurrences of faults and basic equipment failures to the onset of major events like blackouts or massive load sheds require analysis methods and probabilistic tools, which have been developed in reliability theory. The objective of this chapter is therefore to provide the reader with a purpose-driven introduction to reliability and risk analysis concepts, enabling him/her to gain a better understanding about the probabilistic methodology that is developed in detail in the next chapters of this book.

The material presented below starts from a system approach, and goes down to component reliability aspects. The classical Boolean elements of nuclear PSA studies are presented and commented based on electrical examples. Doing so, both qualitative and quantitative analysis methods are explained.

In the last part of the chapter, the implicit assumptions of the Boolean PSA methodology are discussed, in order to highlight the limitations of this approach. Elements of dynamic reliability, which enables a conceptually sound modelling of accident sequence delineation and quantification, are finally provided.

4.2 From Risk to Probabilistic Safety Analysis

4.2.1 Risk

In the context of this book, our interest will focus on the classic definition of technological risk, and its subsequent mathematical expression that is the basis for its evaluation.

L. Haarla et al., *Transmission Grid Security*, Power Systems,
DOI: 10.1007/978-0-85729-145-5_4, © Springer-Verlag London Limited 2011

The notion of risk is two-fold [1]. First, risk relates to the existence of a hazard, of a potential source for damage associated to an activity, as usually stated in dictionaries. For instance, risk is defined as "the possibility of something bad happening at some time in the future; a situation that could be dangerous or have a bad result". The second ingredient of risk consists in the actual occurrence of the events causing this hazard to develop in actual consequences either on people, the environment or the technical system.

Therefore, in order to incorporate both the magnitude of the consequences of an event and the likelihood of its occurrence, risk, interpreted as a measure, can be defined as "the combination of the probability of occurrence of some harm and the severity of that harm" (adapted from IEC 60050-191 [2]).

In mathematical terms, this leads to defining the risk r of a given damage as the product of its magnitude m times its occurrence frequency f

$$r = f \cdot m. \tag{4.1}$$

In many cases, the feared event is likely to occur after initial perturbation develops according to various sequences of events, or scenarios. If each scenario i occurs with a given frequency f_i and entails consequences m_i, then risk is expressed as

$$r = \sum_{i=1}^{s} f_i \cdot m_i, \tag{4.2}$$

where s is the total number of scenarios leading to the damage under study.

In order to account for the perceived risk, and the fear of high-consequence events, a variation of this formula was proposed (e.g. Zio [4]):

$$r = f \cdot m^k, \tag{4.3}$$

where $k > 1$ is an exponent embodying the importance given to consequences with respect to the probability of the damage. Though k can be seen as a subjective parameter dependent on the perception of each individual. It can also be used to include when the regulation is based on a maximum value of risk, the following principle: the higher the consequences of an event, the lower its frequency of occurrence should be, and the acceptable frequency should decrease faster than the increase in damage magnitude. Other descriptions of perceived risk rest on an equiprobable consideration of all damage scenarios, possibly with a probabilistic cut-off on the scenarios taken into account.

In the following, definitions given in Eqs. 4.1 and 4.2 will usually be considered. Anyway, no matter what definition above is given the allowance to, risk is always made up of the following three fundamental components: scenario, frequency and consequence.

4.2.2 *Potential Risk Versus Residual Risk*

Potentially harmful industrial activities, such as chemical or nuclear power plants because of toxic or radioactive releases, respectively, do present a potential risk for the population and the environment due to their exploitation. Yet these plants are of course operated with a set of protection systems and procedures, designed to reduce the actual risk down to a much lower value. The efficiency of these safeguards plays a crucial role in the acceptance process of these activities, and it is the utility's responsibility to make the proof of it for the regulatory bodies. Though the consequences of a blackout are obviously less dreadful than those mentioned above, the grid structure and operation are such that the risk of a blackout or large amount of energy not served to customers is expected to be significantly reduced. This is mainly achieved by a sufficient level of redundancy obtained by the meshed grid, by the satisfaction of the $N - 1$ criterion in all operational circumstances, by efficient defence plans and by adequate asset management and maintenance operations to keep the infrastructure as reliable as reasonably achievable.

As discussed in Chapt. 2, however, a priori insignificant situations (in terms of probability), not covered by the protection policy of the power system, are likely to develop in accidental transient with high consequences. The corresponding risk is called the residual risk, given all measures taken to protect the grid against the occurrence of undesired events. This residual risk should be correctly evaluated in order to assess the relevance of additional adequate mitigative actions and systems.

4.2.3 *Probabilistic Safety Analysis*

Be it referring to safety (for nuclear power plants) or security (for electrical grids), PSA actually aims at identifying the contributors to the residual risk of the activity under study and at evaluating it, in order to provide the utility or system operator, respectively, with priorities for proper cost-efficient, safety/security-driven action.

Consider again Eq. 4.2 giving the expression of the residual risk. Its evaluation rests on the identification of 'all' (or more realistically all significant) scenarios leading to damage. This first asks for the prior definition of a list of credible (combinations of) perturbations (that will often be referred to as *initiating events* or *initiators* in PSA terms, but more generally called *contingencies* for power systems) challenging the system safety or security, i.e. triggering a transient from either a steady-state condition, or a balanced situation between generation and load. Once the initiators are defined and their frequencies assessed, a proper delineation of all damage scenarios that can be reached following the occurrence of one of these initiators must be completed. The conditional probability of following a given sequence, given the initiator occurrence, must be assessed to provide the scenario frequency. Finally, a consequence analysis must be carried out along the scenario to evaluate the damage magnitude that is caused to the system.

Therefore, the PSA study must be completed by following these four steps:

1. the definition of the initiators to be considered, and estimation of their occurrence frequency,
2. the delineation of all possible accident scenarios triggered by a given initiator, i.e. all possible ordered sequences of events starting from this initiator to a failed state,
3. the estimation of the conditional probability of this scenario, and
4. the evaluation of the magnitude of the consequences along this scenario.

All these elements allow to computing the residual risk of the system.

However, it should be emphasized that the actual value of r is not the most interesting part of the study. First, the intimate knowledge of the system gained throughout the procedure is already likely to highlight potential vulnerabilities of the system and changes to be brought. Second, the probabilistic weight associated with all sequences allows to building up an importance ranking between the scenarios, as to their respective contribution to the residual risk. As budgets are always limited, this ranking underlines which contributors to risk should be reduced. How cost-efficient an additional safety measure can be for these highly contributing sequences has then to be estimated in order to adequately invest the available budget in safety improvement.

4.3 Boolean PSA Methodology

4.3.1 From an Initiator Occurrence to a Failure State: Event Trees

Assuming that all initiating events worth being analysed have been identified, a first issue consists in connecting this occurrence of such an event to the possible final consequences it can entail on the system. This is classically done via inductive logic where a branching structure, called an *event tree* [3], is built from the initiator occurrence on. Events likely to take place after the start of the transient are then considered in an ordered fashion. The output of each event, generally expressed in terms of failure and success, gives rise to the split of the current branch of the tree into two branches, and so on.

This concept is illustrated in Fig. 4.1, where a typical event tree is presented. After the occurrence of the initiating event, a first safety function is solicited. It can either succeed or fail, and consequently two branches are created for each of the possible results. Additional safety functions can be activated, successfully or not, to extend the branching process, until a last safety barrier is challenged and preserves its integrity or not.

To each branch event are associated a success probability and the complementary failure probability. The probability of a scenario is obtained by multiplying the probabilities along the corresponding branches. As dependences

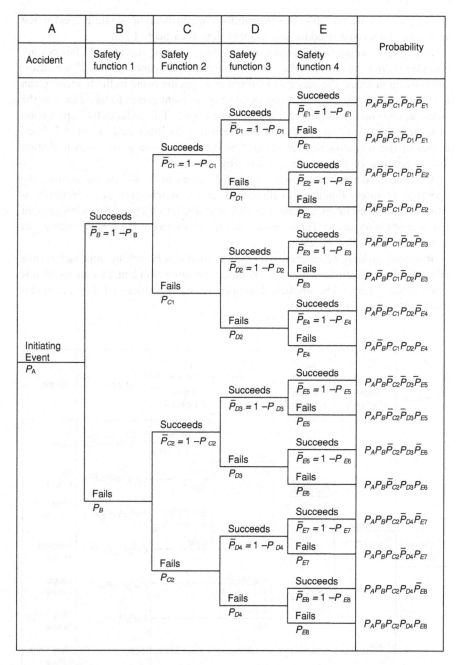

Fig. 4.1 An example of event tree (adapted from Kumamoto and Henley [3])

between events are possible, this probabilistic quantification should be performed with care. This issue is addressed in more details in Sect. 4.3.5.

If n binary events (i.e. resulting in only two outputs, success and failure) are considered after the initiator, the event tree contains 2^n branches, thus 2^n scenarios. Actually, in practice, the success or failure of a specific event in the branching can technically or physically hinder one or more subsequent events to take place, as the latter depends on the fulfilment of the anterior events. This induces the suppression of some branches, i.e. a *pruning* of the event tree, as illustrated in Fig. 4.2. Each output of the scenarios deduced from this inductive construction is then characterised in terms of *final states* reachable by the system.

Events considered in the construction of the tree can be of different natures: the operation of safety functions as illustrated above, intervention of protection systems, fulfilment of safety criteria, the integrity of safety barriers, the occurrence of physical phenomena etc. These events, mentioned on top of the corresponding part of the tree, are called *headers*.

It should strongly be emphasised (Zio [4]) that this branching approach results from the discretization of the real, dynamic, possible accident evolutions in few macroscopic events. The implicit assumptions and limitations of this are further discussed in Sect. 4.4.

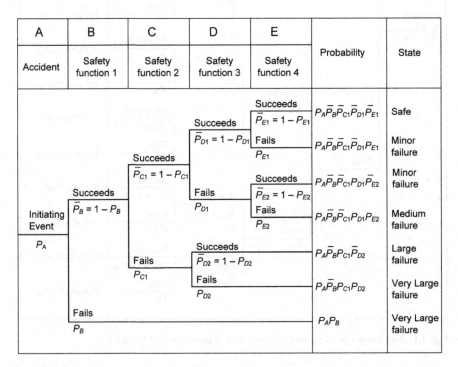

Fig. 4.2 A simplified version of the event tree in Fig. 4.1, after pruning impossible branches (adapted from Kumamoto and Henley [3])

The construction of event trees associated with a substation is developed in detail in Chapt. 6.

4.3.2 From a Subsystem Failure Down to Basic Component Failures: Fault Trees

The events appearing in the headers of an event tree are usually macroscopic, as stated here above. The evaluation of the success or failure conditions constitutes as such another engineering problem that the reliability analyst must tackle. Moreover, a direct quantification of the header probabilities is (almost) impossible, and at least statistically not recommended. Indeed, headers are in some cases one-of-a-kind safety functions, with a unique arrangement of simpler items. In other cases, though the corresponding systems are replicated several times in the whole infrastructure (a typical case for a transmission grid), these systems are highly reliable. Records of failures are scarce and do not allow a direct, accurate estimation of their occurrence probabilities. On the contrary, failure data corresponding to more basic system elements are more easily collected and available in a statistically relevant way to perform probability estimation. The occurrence of these *basic events* must then be connected to that of the macroscopic events, which will be referred to as *top events* in the following.

This is performed by way of a systematic, deductive analysis of the root causes of these top events, via another branch structure called a *fault tree*. A top event is decomposed in a vertical arrangement of either the combination (logical intersection—AND operator) or competition (logical union—OR operator) of less complex situations, called *intermediate events*. The latter are in turn analysed in their lower-level intermediate events, and the process goes on until the entries of the logical operators AND and OR are basic events, or events that the analyst does not wish to further decompose.

Figure 4.3 provides some simple examples of decomposition of intermediate events in their causes, using the OR and AND logical operators, respectively. The lack of circuit breaker trip is caused by the competition between the circuit breaker failure and the lack of trip signal, any of these two events causing the occurrence of the upper event. The 'No trip signal' event is then caused by the combined failure of two relays (to send a trip signal), none of these lower events being able alone to make the upper event true.

Logical relationships like union and intersection between events constitute the *gates* of the fault tree. Figure 4.4 provides two alternative representations of the most common AND and OR gates. Fault tree construction often rests on some additional gates that can facilitate the analysis work of the top event, like e.g. the priority AND gate (when all input events have to occur in the order from left to right), the exclusive OR gate (when one, but not all, input events occur), or the k-out-of-n voting gate. It should, however, be stated that these additional gates all

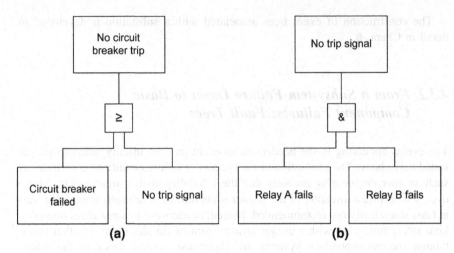

Fig. 4.3 Two fault trees **a** presents an example of OR gate and **b** an example of AND gate (adapted from Zio [4])

Fig. 4.4 Gates in a fault tree [3]

Gate symbol IEE	Gate symbol ANSI	Gate name	Causal relation
&		AND gate	Output event occurs if all the input events occur
≥1		OR gate	Output event occurs if any of input events occurs
=1		XOR gate	Output event occurs if one, but not both, of the input events occur

reduce to combinations of AND and OR gates [3]. The top event decomposition is therefore driven by Boolean logic.

The graphical representation of events is provided in Fig. 4.5. Basic events, sometimes called the *leaves* of the fault tree, correspond to circles. Diamonds refer to *undeveloped events*. These are usually intermediate events, whose further decomposition is not considered as relevant for the analysis. A circuit breaker failure could be seen as one of these undeveloped events, the internal causes of this failure corresponding to too many details in a grid security analysis. On the contrary, a rectangle corresponds to an intermediate event that still has to be developed.

All events are random by nature. However, a *house event* can be substituted to a basic event to force its occurrence (or non-occurrence). This allows to analysing in the quantification of the top event probability (see Sect. 4.3.5) the sensitivity of the results to individual basic events.

Fig. 4.5 Events in a fault
tree [3]

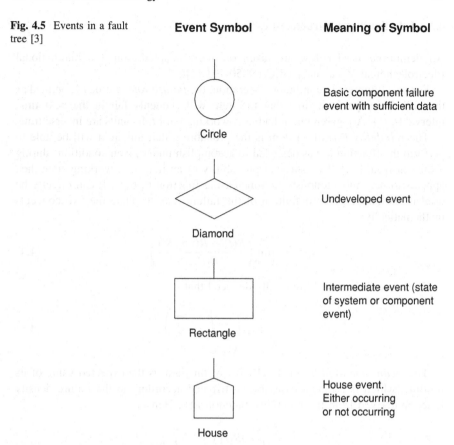

Event Symbol	**Meaning of Symbol**
Circle	Basic component failure event with sufficient data
Diamond	Undeveloped event
Rectangle	Intermediate event (state of system or component event)
House	House event. Either occurring or not occurring

There is no unique way of building a fault tree for a complex top event. The exhaustivity of the decomposition is analyst-dependent. Yet heuristic recipes to help the reliability analyst in his/her task can be found in Kumamoto and Henley [3] or in Zio [4].

4.3.3 Characterising Basic Events: Component Reliability

At this stage, events corresponding to the headers of an event tree have been decomposed through a fault tree in their logical combinations of basic events. The latter are the elementary bricks of the whole construction. They correspond to events occurring frequently enough to associate probabilistic figures with their occurrence.

The present section hence describes basics of component reliability for the characterisation of basic events appearing in fault tree analysis.

4.3.3.1 Reliability Characteristics

All definitions used below are taken or freely adapted from the International Electrotechnical Vocabulary (IEC 60050-191 [2]).

For an equipment in continuous operation, the *failure rate* at time t is defined as the probability per unit time that the item will suddenly fail in the next time interval $[t, t + \Delta t]$, given that it had kept working until t. Its units are inverse time.

The *reliability R(t)* of an item is the probability that this item will be able to perform the function it was designed to accomplish under given conditions during a time interval $[0,t]$. This survival probability of an item in a working state then appears to be a monotonously non-increasing function of time. It can directly be used in the above given definition of the failure rate, to relate the two concepts mathematically:

$$\lambda(t) = \lim_{\Delta t \to 0} \left(\frac{R(t) - R(t + \Delta t)}{R(t)\Delta t} \right). \tag{4.4}$$

Therefore, it is straightforwardly deduced that

$$R(t) = \exp\left(-\int_0^t \lambda(s)ds \right). \tag{4.5}$$

The *mean time to failure* (MTTF) T_{TF} of this item is the expected value of its lifetime, which is a random variable distributed according to the failure density function derived from the reliability function $R(t)$. Hence,

$$T_{TF} = \int_0^\infty t\left(-\frac{dR(t)}{dt} \right)dt = \int_0^\infty R(t)dt \tag{4.6}$$

where the last part of the above relation being true under general conditions. For a constant failure rate, the component lifetime obeys a negative exponential distribution of parameter λ, and $T_{TF} = 1/\lambda$. The *failure density function f(t)* of the item, defined as

$$f(t) = -\frac{dR(t)}{dt} \tag{4.7}$$

reduces to the negative exponential probability law.

For components required to work on demand, as circuit breakers for instance, their likelihood to fail can be modelled in two ways. First, a failure rate (per unit time) describes its time-continuous proneness to fail, even if failure is usually detect only when a fault is not eliminated or when an inspection takes place. An alternative option consists in using a failure rate per demand (or a failure probability) to model the probability that the circuit breaker does not work as expected

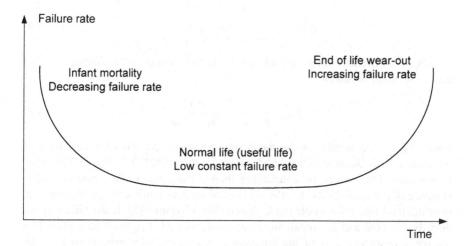

Fig. 4.6   Failure rate evolution with time—the bathtub curve

when asked to open. In the sequel of this book, failure rates per unit time are used for such items.

Failure rates are not constant throughout the equipment lifetime. In analogy with the death rate of a human being, feedback from field data shows that failure rates of technical items usually follow a bathtub-shaped evolution in time (see Fig. 4.6).

The first zone in Fig. 4.6 corresponds to the infant mortality of the component, where early failures can take place due to design or manufacturing defects, for instance. The second zone is characteristic of the useful life of the item, during which failure occurs at random and $\lambda(t)$ can be seen as approximately constant. Finally, the wear-out zone describes the effect of ageing on the failure rate, which tends to significantly increase.

Availability is defined as the ability of an item to be in a state to perform a required function under given conditions at a given instant of time or over a given time interval, assuming that the required external resources are provided. Availability does no longer correspond to a survival probability as seen for reliability (except for a non-repairable item), but to the probability of existence of a given working performance of the item at a given time. As reinforced by the International Electrotechnical Commission (IEC 60050-191 [2]), this ability depends on the combined aspects of the reliability performance, the maintainability performance and the maintenance support performance.

Let us define the repair rate $\mu(t)$ of an item, in analogy with the failure rate defined previously. Though it is hardly defendable that repairs could take place at random, the approximation of a constant μ is often taken. In this case, the availability $A(t)$ of a basic, binary component, being either working or failed, and subject to perfect repairs bringing it back to as-good-as-new conditions, writes

$$A(t) = \frac{\mu}{\lambda + \mu} + \frac{\lambda}{\lambda + \mu} e^{-(\lambda+\mu)t}. \tag{4.8}$$

After a sufficiently long time of operation, the asymptotic availability A_∞ of this component becomes

$$A_\infty = \frac{\mu}{\lambda + \mu} = \frac{T_{TF}}{T_{TF} + T_{TR}} \tag{4.9}$$

where $T_{TR} = 1/\mu$ stands for the mean time to repair (MTTR), the definition of which is directly adapted in the failed state from that of MTTF in the working state of the item. The asymptotic availability then corresponds to the intuitive understanding of the availability, i.e. the fraction of time a component spends working, over the total time of a whole cycle "operation + repair" [5]. If the failure is not directly detected and the repair instantaneously started, T_{TR} must be replaced by the mean downtime T_{DT} of the component, which includes aspects such as the detection time, the waiting time before repair, the time to restart the item.

4.3.3.2 Failure Mode Identification

A crucial aspect of a system reliability analysis consists in properly identifying, both at the component and system levels, all ways in which items can fail and the consequences of these failures on other items of the system.

This operation is usually carried out in the failure modes and effects analysis (FMEA). It consists in constructing a table (Table 4.1) where all items of the system are systematically listed in the first column, possibly with their reference numbers. The function of each item is then provided. The different modes in which it can fail are listed, as well as the possible cause(s) of this failure mode and the effects it can bring to other parts of the whole system. Detection means are associated with each failure mode.

Usually, some quantitative or semi-quantitative information are additionally provided. The example of Table 4.1 is extended in Tables 5.2 and 5.3, by providing the failure rates and unavailability values corresponding to the different items. In other cases, when the FMEA table aims to be used as a first decision-aiding tool before performing a full probabilistic analysis, a risk index is estimated for each failure mode, in order to prioritise possible interventions in the system. This risk index results from expert elicitation, on integer scales, of the likelihood of occurrence of the failure under study and the expected seriousness of the consequences. The classical factors of risk are thus approximated in this way, before completing an actual probabilistic risk analysis for the system.

Practical rules to efficiently fill in a FMEA chart fall beyond the scope of this book. The interested reader can refer to McDermott et al. [6] or Rausand and Høyland [7] for further information on this topic, which brings insight in the system vulnerabilities before any qualitative or quantitative full-scale security analysis of the system is conducted.

Table 4.1 Excerpt from a FMEA table, failure mode and effect analysis for a microprocessor distance relay [8]

Item type	Function	Cause of failure	Failure mode	Effect of failure	Detection of failure
Processor distance relay	Send a trip signal during faults	Software error	No trip signal	No trip signal	Test
Processor distance relay		Human error	Current measurement circuit of the relay is disconnected	No trip signal	Test
Processor distance relay		Human error	Terminal strips of the relay or of the relay cubicle are disconnected	No trip signal	Test
Processor distance relay		Human error	Erroneous setting or configuration	No trip signal	Test
Processor distance relay		Ageing	The spring of the card joints is loose, the signal transfer is prevented	No trip signal	Test
Processor distance relay		Ageing of the components in the relay power supply	Relay loses the power supply	No trip signal	Alarm
Voltage measurement circuit	Get voltage measurement to the relay	Human error: voltage measurement circuit disconnected	No trip signal due to voltage transformer supervision	No trip signal	Alarm

4.3.3.3 Failure Rate Estimation

Failure rates are far from being universal constants, even when they are considered only in the useful lifetime of items. They are classically estimated via statistical inference from the results of test campaigns and/or from field data, the latter being less expensive and closer to the actual operational conditions of the system the item lies in.

One of the most used estimation techniques is the maximum likelihood estimation (MLE) method [9]. A probability law for the failure density function of a given item is first selected. From failure and operation data obtained from a set of identical items, the likelihood function of this sample is defined as the probability (or probability per unit time) to observe it. The MLE method consists in choosing as estimator of the failure rate (or of the different parameters of the failure density function) the value maximising the likelihood of the sample at hand. The MLE method can treat both failure and censored data, and the resulting estimator displays interesting statistical properties. A confidence interval can be estimated to define an interval about the estimated value, where the failure rate lies with a predefined probability level.

Confidence intervals can be very large if the sample population is limited, however. In such a case, the maximum likelihood estimator can become of limited interest, given the uncertainty on its actual value. This is because the MLE method derives from the classical frequency-based understanding of probability, resting on the availability of a large number of random trials. This is obviously not true when failure data are scarce, or when the set of items considered in the sample display some differences or when they are not operated in fully identical conditions. Even in the sample of failure data is large enough and was collected in identical conditions, what if the item, whose failure rate is estimated, is incorporated in a new system or operated in a new environment? The implicit assumption of frequency-based probabilities, according to which the future random trials will be identical to those completed in the past and on which probabilities were estimated, does no longer hold.

A way to circumvent this difficulty is to resort to Bayesian estimation [10], based on the Bayes theorem, well-known in probability theory. Consider an event A, and a complete set of events H_i, $i = 1 \ldots n$. This means that any couple of events (H_i, H_j) are mutually exclusive $(P(H_i \cap H_j) = 0)$ and that $P(H_1 \cup H_2 \cup \ldots \cup H_n) = 1$. Then, it can easily be shown that

$$P(H_i|A) = \frac{P(A|H_i) \cdot P(H_i)}{\sum_{j=1}^{n} P(A|H_j) \cdot P(H_j)}, i = 1 \ldots n \qquad (4.10)$$

To highlight how this mathematical theorem is used for estimation, let us first rewrite it in a continuous form

$$f(\lambda|t_1 \ldots t_n) = \frac{L(t_1 \ldots t_n|\lambda) \cdot f_0(\lambda)}{\int L(t_1 \ldots t_n|\lambda') f_0(\lambda') \mathrm{d}\lambda'} \qquad (4.11)$$

where $t_1 \ldots t_n$ are a sample of failure times of the item under study, observed during its current operation, and λ stands for its failure rate (hence implicitly assuming a negative exponential law for the failure density function of the item). In the above expression, $f_0(\lambda)$, named *prior distribution*, corresponds to a probabilistic knowledge of the value of the failure rate, prior to obtaining any field data in the current operational context. This prior distribution can be based on a database for similar items or for the same unit in a different context, or obtained by expert judgement. $L(t_1 \ldots t_n | \lambda)$ expresses the likelihood of the sample of failure times, i.e. it is proportional to the probability to observe the sample of failure data at hand, given λ. This likelihood is the function to be maximised in the MLE method mentioned above. In the Bayesian estimation, however, the likelihood function is weighted by the prior distribution to provide, after normalisation, a *posterior distribution* of λ given the observed sample, i.e. $f(\lambda | t_1 \ldots t_n)$.

The Bayesian approach thus allows, when statistical data are limited, to including in the estimation of the parameter some available expertise to enrich the field data. The *Bayesian estimator* of λ is given by the expected value of the posterior distribution.

Prior distributions depend on the type of prior information available. As a particular case, in non-informative cases, i.e. when the prior knowledge about the failure rate is limited to a given interval, a uniform probability law is classically selected.

4.3.3.4 Common Cause Failures

Basic events in fault trees and event trees are not always independent of each other. The same cause (e.g. flood, fire, explosion, erroneous manoeuvre, earthquake) can provoke the simultaneous failure of more than one item in a cascade that has to be considered as one event. In other circumstances, one event alone affects (and usually increases) the failure probabilities of some set of elements, simultaneously and all in the same way. Such an event can be the occurrence of more challenging operation conditions for different components. It can also be linked to errors in assembling, or it can be due to repeated errors in the tuning of detection or protection devices, to name some possibilities among numerous cases.

In the fault tree structure, the existence of common cause failures (CCF) can be accounted for in the following way. A basic event corresponding to the failure of a component belonging to a CCF group is replaced with an OR gate, the entries of which are the independent failure mode of this component as well as all other basic events implying its failure together simultaneously with the failures of other components of this group.

A simple way of modelling the dependence between component failure probabilities is the β factor approach [11]. Assume a set of m identical components is subject either to independent failures or to CCF. Parameter β then embodies the conditional probability of having the m components failing simultaneously, given one component fails. A zero probability is set to other basic common cause events with k simultaneous failures, if $1 < k < m$. This simple approach is therefore conservative, the probability of all intermediate situations being included in that of the worst case. The extensions of this model were proposed (e.g. multiple Greek letters, α factors) to release the latter assumption, but their presentation falls beyond the scope of this section.

Besides the probabilistic modelling of the common cause failures, reliability and security analysts should pay a particular attention to various situations likely to introduce unexpected dependences between failure events. An important common cause affecting transmission grid security can also be found in lack of maintenance of vegetation below lines, likely to cause earth faults when the line sag increases. This situation was contributing to the North-East American blackout in 2003, for which the sequence of events comprises several line faults due to this (geographically distributed) lack of prevention.

Another example of unusual common cause failures comes from thermal effects in overloaded lines, after the occurrence of a first fault. Abnormal values of temperature can cause an increase of the failure rates of the lines by one or two orders of magnitude. Such a situation could entail multiple line trippings at the start of a blackout.

Such situations either create new sequences that were overlooked in the event tree analysis, or bring the probability of identified accident sequences, which were first assessed as negligible, to significant values. They should therefore be analysed with much care and systematically tracked.

4.3.4 Identifying Vulnerabilities: Minimal Cut Sets

In Sects. 4.3.1 and 4.3.2, event trees were shown to describe the possible scenarios following the occurrence of an initiating event, while fault trees were introduced to analyse an event at the level of a subsystem in terms of logical combinations of basic events at the level of components. Though the concepts behind this combined approach are quite simple, the trees obtained in the analysis of a real-size system can soon become large, and consequently their interpretation difficult.

In order to summarise the information contained in a fault tree in a practical way and perform a qualitative analysis of the system vulnerabilities highlighted by the logical decomposition of the top event, the concept of minimal cut set is introduced [3]. A cut set is a set of basic events, the occurrence of which causes the top event to take place. A minimal cut set (MCS) is a cut set containing no other cut set. In other words, a minimal cut set contains the minimal number of basic events that have to be realised at the same time to provoke the system failure.

The *order* of a minimal cut set is defined by the number of basic events contained in this cut set. Before associating any probabilistic quantification to these events, it is directly observed that minimal cut sets of lower orders correspond to more challenging system vulnerabilities than high-order minimal cut sets.

Fig. 4.7 Equivalent fault tree based on the minimal cut sets

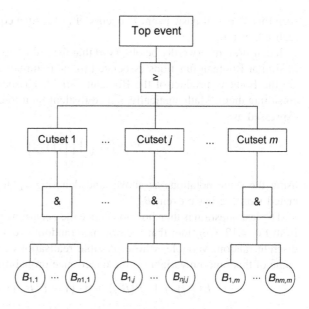

Algorithms for the automatic identification of all minimal cut sets of a fault tree can be found in the literature [3]. They rest on the following simple observations: for a fault tree made of a unique OR gate, all entries of this gate are minimal cut sets of order 1, while for a fault tree made of a unique AND gate, there is only one minimal cut set containing all entries of the gate. When analysing a more complex fault tree, minimal cut sets can be obtained in a top-down approach, considering that each OR gate increases the number of minimal cut sets while each AND gate increases the number of events in a minimal cut set.

As any of the minimal cut sets causes the top event to occur, and as a minimal cut set is realised only if all its events occur together, an equivalent fault tree can be drawn, as displayed in Fig. 4.7. In this figure, the leave $B_{i,j}$ of this equivalent fault tree is the ith basic event belonging to the jth minimal cut sets of order n_j, $j = 1...m$. This representation will be useful in the next section.

4.3.5 Quantifying Fault Trees and Event Trees

The fault tree/event tree logical decomposition performed so far has addressed the first requirement of a probabilistic safety analysis, i.e. the identification of the accident scenarios, and has partly contributed to the assessment of the consequences of these scenarios. Yet the issue of quantifying the probability of each sequence of events challenging the safety of the system remains open.

In order to tackle this problem, a *failure structure function* $\Psi(\bar{Y})$ associated with a top event is defined. This failure function is a Boolean function, whose arguments $\bar{Y} = (Y_1...Y_n)$ are Boolean variables related to the n basic events and such that $Y_i = 1$ if basic event B_i occurs. $\Psi(\bar{Y})$ is then equal to 1 if the top event is realised, else 0.

It can straightforwardly be observed that the failure function is the Boolean sum of similar Boolean functions associated to the minimal cut sets, and that the latter are the Boolean product of the Boolean variables associated with the events that constitute them. Mathematically, the equivalent fault tree drawn in Fig. 4.7 can be expressed as

$$\Psi(\bar{Y}) = \bigcup_{j=1}^{m}\left(\bigcap_{i=1}^{n_i} Y_{i,j}\right) \tag{4.12}$$

using the same notations as above, and denoting $Y_{i,j}$ the Boolean variable associated with the basic event $B_{i,j}$.

The last question is then how to deduce the occurrence probability of a top event from Eq. 4.12. Consider that each $B_{i,j}$ is a random event, $Y_{i,j}$ can also be seen as a discrete random variable, whose possible realisations are 0 and 1. The expected value of this random variable is the occurrence probability of event $B_{i,j}$, as

$$E(Y_{i,j}) = 1 \cdot P(B_{i,j}) + 0 \cdot \left(1 - P(B_{i,j})\right) = P(B_{i,j}) \tag{4.13}$$

In the same way, the failure function can be interpreted as a discrete random variable, and we straightforwardly deduced that

$$E(\Psi(\bar{Y})) = P(\text{top event}) \tag{4.14}$$

Then, using the arithmetic equivalence of the Boolean operations, i.e.

$$\begin{aligned} Y_i \cup Y_j &\rightarrow Y_i + Y_j - Y_i \cdot Y_j, \\ Y_i \cap Y_j &\rightarrow Y_i \cdot Y_j. \end{aligned} \tag{4.15}$$

Equation 4.12 can be expanded in an arithmetic expression, whose expected value can be directly taken, according to Eq. 4.14, to provide the top event probability. If all basic events $B_{i,j}$ are independent, the top event probability is expressed as a combination of the $P(B_{i,j})$. In case of dependences between events, models for the probability of the intersection of events have to be introduced, such as those mentioned in the CCF section.

If the different top events appearing along a scenario in an event tree are independent, the scenario probability is directly obtained by multiplying the different top event probabilities, obtained as explained here above. However, if some basic events appear in the decomposition of different top events along a sequence, a Boolean *scenario function* has to be defined by taking the Boolean intersection of the different failure functions for the top events. The resulting complex expression can be reduced using the rules of Boole's algebra. The expected value of the scenario function, interpreted as before as a discrete random variable, provides the probability of the occurrence of the scenario under study. In practice, a probabilistic cut-off is often introduced to prune the event tree from scenarios with a nonsignificant probability, in order to focus the safety analysis on the sequence of events contributing more to the residual risk of system failure.

4.3.6 Carrying Out Sensitivity Analysis: Importance Factors

Having calculated a top event probability or a scenario frequency as seen in the previous section, one can wonder which basic events or group of events contribute the most significantly to the system failure. In order to answer such a question, a sensitivity analysis can be carried out. Various sensitivity coefficients, called *importance factors*, can be found in the literature [7].

Some of the most used importance factors are presented in Sect. 8.2.4, for direct use in the application treated. These sensitivity coefficients highlight the relative impact of a basic event occurrence or non-occurrence on a top event probability. As stated in Sect. 4.3.2, forcing a basic event to either occur or not can be done in a fault tree by means of a house event replacing the classical basic event by a deterministic one. This modification can then be propagated in the failure function to evaluate the top event probability conditional to this forced event. Such conditional probabilities are the elementary ingredients of all importance measures in the system.

The calculation of the importance factors is usually performed under the assumption that the system is coherent. A *coherent system* [4] is a system such that the improvement of any component results in an improvement of the whole system, and never causes the system to fail if it was working. Similarly, any decrease of performance of a component will entail the successful working of the system if it was failed. In mathematical terms, the system failure function is a monotonically increasing function of each of its input variables and does not contain any complemented variables (i.e. negation of events).

Therefore, for a coherent system, $\Psi(\bar{Y}) = 1$ if $Y_i = 1$ for all i, $\Psi(\bar{Y}) = 0$ if $Y_i = 0$ for all i, and

$$\Psi(\bar{Y}) \geq \Psi(\bar{X}) \text{ for } \bar{Y} \geq \bar{X} \tag{4.16}$$

4.4 Towards an Integrated PSA Approach

The coupled event tree/fault tree methodology to conduct a probabilistic risk assessment is conceptually simple. As seen above, possible scenarios developing after the occurrence of an initiating event are represented in an inductive branching structure called an event tree, while the occurrence of the events corresponding to the branch points of an event tree are analysed with a deductive branching logic called a fault tree. Boolean algebra is the basic mathematical tool used in the logical decomposition of major events in the unions or intersections of basic events, and behind the subsequent quantification of the occurrence probabilities of the top events of the various fault trees, and then of the scenario frequencies.

Despite its conceptual simplicity, this binary approach of the grid evolution in transient modes could not fully capture the dynamic response of the grid to a major disturbance. This section reviews the implicit assumptions of the classical PSA methodology, discusses the role of grid dynamics in the delineation of scenarios and in their probabilistic quantification and introduces the main concepts of an integrated approach to PSA.

4.4.1 Implicit Assumptions of the Fault Tree/Event Tree Approach

An event tree presents the possible damage scenarios that could take place after the occurrence of an initiator as an ordered sequence of (top) events, hence of failures or successes. Reducing the complexity of the response of a large infrastructure (like a nuclear power plant or a transmission grid) in this way to an initial disturbance in ordered series of binary events rests on implicit assumptions, which are often overlooked by PSA practitioners [12]:

- The output of the analysis is assumed to be unaffected by a change in the order of the headers of the event tree. Should it be the case, the event tree analysis is duplicated to account for meaningful modifications in the sequence order.
- The output of the analysis is assumed to be unaffected by a change in the timing of the events in the sequence. Again, if such a variation in the timing modifies the results of the PSA study, the analyst is assumed to have noticed it, and adapted the current event tree or built additional ones to take this circumstance into consideration.
- The effect of dynamic process variables on the scenario development is limited to an appropriate definition of the top events as a function of thresholds on the values of these variables.

Obviously, these assumptions are not always satisfied in practice. The event tree/fault tree methodology does not provide in itself a way of assessing the quality of its implementation. As a matter of say, the quality of a PSA analysis will be defined by the quality of the PSA analyst, who turns out to be capable of aggregating the branch events with their consequences on the system dynamics to correctly identify the accident sequences [13]. Yet an integrated approach of the system evolution under configuration changes is necessary to have a sound justification of the process of sequence delineation, identification and probabilistic evaluation. Such a treatment of the PSA problem should combine the system dynamics, the transition processes between system configurations, the operator actions if relevant, and time.

4.4.2 The Role of Grid Dynamics in PSA

The estimations of failure rates usually performed on the basis of failure records averaged on periods of observation should sometimes be considered with care. For instance, the power line carrier unavailability is dependent on the state of the line:

failed or not failed, especially if the power line carrier is installed in two phases instead of three. During 1-phase faults, there is always a healthy phase available. The case is different for 3-phase faults. During 3-phase faults, all the phases carry the fault current and it is very probable that the telecommunication signal cannot pass the faulted line. In this case, the constant unavailability of the telecommunication is 1. The failure has a correlation with the state of the grid, therefore constant unavailability is not a correct way of modelling although often used.

If a device fails once during a year, following, e.g. an exceptional incident, should its failure rate be assessed as one per year, or should the reliability engineer instead consider on one hand its conditional failure probability, averaged on the number of occurrences of the exceptional incidents, or on the other hand the yearly frequency of this exceptional event? If the first modelling option is selected, as it is generally done, the failure rate value could be strongly underestimated, if the likelihood of a component failure has to be considered in the course of a transient scenario, when this exceptional incident is likely to be encountered. The second modelling would lead to a more significant probability of the scenario including the component failure, given the occurrence of the initiating event leading to these abnormal operation conditions for the device.

The dynamic evolution of the transmission grid is thus likely to affect the probability of blackout scenarios, hence explaining why some a priori too low-probable situations are sometimes met in practice. Besides this first possible effect on the PSA study, grid dynamics actually appears as the driving force of scenario development after stable conditions are left, following the occurrence of an initiator. Grid dynamics control the redistribution of currents and flows in the lines after the first fault or any subsequent event, it defines the time at which protection settings are exceeded, it makes appear the possible differences between grid evolutions following a given scenario but with variations in the timing of events. In pre-cascade situations, the thermal evolution of overloaded lines after a first line fault affects their proneness to fail in a significant way, as already stated above. This happens with characteristic times much larger than the electrical time constants, and it is usually not included in the security study. Yet this thermal evolution of the grid is likely to induce additional contingencies, hence the start of an accident scenario.

And of course, grid dynamics is an essential element of the consequence analysis completing the PSA study.

4.4.3 Elements of Dynamic PSA

For about two decades, dynamic PSA methodologies have been proposed and developed in the field of nuclear or aeronautics safety. They mainly aim at automating the identification and generation of possible accident scenarios, by correctly accounting for the interaction between the continuous system dynamics and the discrete configuration changes. Indeed, each transition between two

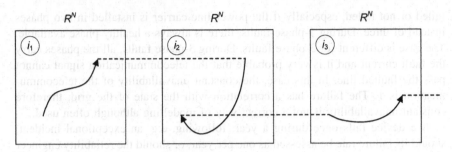

Fig. 4.8 A piecewise-deterministic stochastic process [10]

configurations modifies the dynamic evolution laws of the system, while in turn the system dynamics affects the transitions between configurations [13, 14]. This happens either because the latter are defined by setpoints on the dynamic process variables, or because the failure rate values are affected by these process variables.

The system evolution obeys a piecewise-deterministic stochastic process, schematically represented in Fig. 4.8. Following the occurrence of an initiator, the system leaves its steady-state conditions to enter configuration i_1, where it obeys deterministic evolution equations associated to i_1. The deterministic evolution of the system in this configuration stops as soon as a transition to configuration i_2 takes place, and a new set of evolution equations defines the system evolution. Though this transition from i_1 to i_2 can be a stochastic event, the system evolution in i_2 keeps deterministic, as long as no new event takes place and brings the system in configuration i_3, and so on. These stochastic bifurcations between deterministic evolutions can in some cases drive the system outside a safety region, where damage situations are encountered.

As branching events between system evolutions can take place at any time, this approach was also called the *theory of continuous event trees*. It offers a methodological framework from which approximations can be deduced and controlled, contrarily to the classical event tree/fault tree technique.

Applying such an approach to industrial cases requires extensive computer effort for simulation though efficient Monte Carlo algorithms can reduce the numerical workload of the approach. First attempts at using this integrated methodology to assess grid security can be found in [15, 16].

4.5 Conclusions

This chapter reviewed the essential motivations and ingredients of a probabilistic safety analysis of an industrial system, in order to apply it to the issue of the grid security. The main concepts of the event and fault tree methodology were presented, in order to derive accident scenarios and relate them to the combinations of occurrences of basic events. The last section presented the elements of dynamic PSA methodologies, after listing some implicit assumptions of the classical event tree/fault tree technique and showing how grid dynamics affects the classical PSA process.

In the context of this book, the construction of event trees following a fault line is limited to the response of the different protection levels and telecommunication devices at the ends of the line. In such a case, it is reasonable to decouple the construction of event trees and the dynamic analysis of transients. Sequences of the event tree are constructed classically. Grid dynamics is then used to perform the consequence analysis along the scenarios identified.

References

1. Aven T (2010) Misconceptions of risk. Wiley, Chichester. ISBN 978-0-470-68388-0.
2. IEC 60050-191 International Electrotechnical Vocabulary (IEV), Chapter 191: Dependability and quality of service. http://www.electropedia.org/iev/iev.nsf/index?openform&part=191. Accessed 29 June 2010
3. Kumamoto H, Henley EJ (1996) Probabilistic risk assessment and management for engineers and scientists, 2nd edn. IEEE Press, ISBN 0-7803-6017-6
4. Zio E (2007) An introduction to the basics of reliability and risk analysis. Ser on qual, reliab and Eng stat, 13. World Scientific Publishing Co Pte Ltd, Singapore ISBN-13 978-981-270-639-3
5. Bedford T, Cooke R (2001) Probabilistic risk analysis—foundations and methods. Cambridge University Press 2001, Cambridge
6. McDermott RE, Mikulak RJ, Beauregard MR (1996) The basics of FMEA. Quality Resources, New York
7. Rausand M, Høyland A (2004) System reliability theory, models, statistical methods, and applications, 2nd ed. Wiley, Hoboken, NJ, USA. ISBN 0-471-47133-X
8. Pottonen L (2005) A method for the probabilistic security analysis of transmission grids. A doctoral dissertation, Helsinki University of Technology, 951-22-7591-0, 951-22-7592-9 http://lib.tkk.fi/Diss/2005/isbn9512275929/. Accessed 29 June 2010
9. Misra KB (1992) Reliability analysis and prediction—a methodology oriented treatment. Fundamental studies in engineering 15. Elsevier, Amsterdam
10. Lannoy A, Procaccia H (1994) Méthodes Avancées d'Analyse des Bases de Données du Retour d'Expérience Industriel. Collection de la Direction des Etudes et Recherches d'Electricité de France, 86 Eyrolles, Paris
11. Modarres M (1993) What every engineer should know about reliability and risk analysis. Marcel Dekker Inc., New York
12. Siu N (1994) Risk assessment for dynamic systems: an overview. Reliab Eng Syst Saf 43(1):43–73
13. Labeau PE, Smidts C, Swaminathan S (2000) Dynamic reliability: towards an integrated platform for probabilistic risk assessment. Reliab Eng Syst Saf 68(3):219–254
14. Labeau PE, Izquierdo JM (2005) Modeling PSA problems. I. The stimulus-driven theory of probabilistic dynamics. Nucl Sci Eng 150(2):115–139
15. Chen Q (2004) The probability, identification, and prevention of rare events in power systems. PhD dissertation, Iowa State University, Ames, Iowa. http://www.pserc.org/cgi-pserc/getbig/publicatio/2004public/qimingchen_phd_dissertation_on_cascading.pdf. Accessed 13 May 2010
16. Hortal J, Izquierdo JM (1996) Application of the integrated safety assessment methodology to the protection of electric systems. Reliab Eng Syst Saf 52(3):315–326

Chapter 5
Grid Faults and Component Failures

5.1 Introduction

The chapter describes grid faults; mainly short circuits and trips of big generators. These common faults or credible events affect system stability and security. The chapter also discusses component failures, their causes and the calculation of failure rate estimates and unavailabilities. Statistics are briefly described, too. The focus in all these subjects is in the grid security and in the substation reliability model.

Section 5.2 handles shunt faults, generator trips, and protection reliability. Faults, when connected to protection misoperations (missing or unselective trips), can create major disturbances or blackouts. Both missing trips and unselective trips can be risky for the system security. A missing trip by the main protection relays leads to long fault duration even though the backup protection would operate correctly. Backup protection systems have longer delays than the main protection before sending a trip command. Often they also trip more components than just the faulty one.

Section 5.3 describes component failures focusing on the most important components needed in the PSA substation model: circuit breakers, protection relays, and telecommunication channels. Pottonen [1] presents an example of component failures for the substation model, derived from the Finnish 400-kV component failure statistics. Section 5.4 gives some thoughts of statistics and presents some line fault statistics.

5.2 Shunt Faults

5.2.1 System Effects of Shunt Faults

Faults are stochastic events and their average frequencies vary in different countries and regions. Climate, geography, and line types, for example, affect the fault

L. Haarla et al., *Transmission Grid Security*, Power Systems,
DOI: 10.1007/978-0-85729-145-5_5, © Springer-Verlag London Limited 2011

frequency. Power system faults are either shunt or series faults. Short circuits and earth faults are shunt faults, where the fault current flows between the phase conductors or between the phase(s) and earth. Lightning strokes, for example, causes shunt faults. A series fault is a fault for which the impedances of each of the three phases are not equal, usually caused by the interruption of one or two phases.

Short circuits cause currents that usually are bigger than load currents. (If a fault occurs at the remote end of a long radial line, the fault current can be smaller than the maximum load current). Earth fault currents are high, too, if the system is earthed. Fault currents can damage the components due to extra heat, cause fire, instability or even sagging of lines due to thermal expansion. During earth faults, the currents flowing in the soil can cause touch and step voltages that may be dangerous to people and animals. Due to these reasons, the faulty part of the grid needs to be disconnected from the healthy parts. The function of the protection systems is to detect the fault, send a trip command to the circuit breaker(s), which disconnect the faulty part from the grid. After the damaged part is disconnected, the healthy parts of the grid can continue to transmit power.

The power system state after the fault depends on the type of fault, the fault location, fault duration, and the power system state before the fault.

When the power system is operated according to the $N - 1$ principle, it survives a credible fault and the following trip of a component provided that the protection operates correctly. If the angle stability of the system sets the limits for power transmission, there is a critical fault duration. The duration of a fault should always be shorter than the critical fault duration; otherwise, the system or a single generator may lose its stability. If the thermal limits of the lines are the limiting factor, security may be lost if there are one or more unselective trips in addition to the correct trip.

If the dynamic instability of the system sets the limits for the power transmission, the duration and location of the fault are significant. Too long fault durations (especially near big generators) may lead to instability, while with normal fault durations the grid remains stable. In system planning, power flows and protection specifications (that influence the fault durations) are coordinated. In other words, the normal fault clearing times define the maximum power flows that are secure. If a missing trip occurs (the main protection systems fail to function), the fault duration is longer, even if the backup protection acted as planned. The redundant main protection systems reduce the risk of missing trips.

Usually, the trip times of protection systems are the same in all substations even though, in principle, the grid can withstand shunt faults with longer durations far away from the big generators than near them. It would be difficult to plan a selective protection for the whole grid, if the tripping times vary according to the location.

The synchronous generator's shafts accelerate during nearby faults, because the faulty lines cannot transmit power from the turbine to the grid. If the voltages are low, the electrical power is reduced as Eq. (3.1) presents it. If the fault duration is too long, the shaft accelerates too much during the fault. In this case, the shaft cannot reach the synchronous speed after the fault is over, the generator loses its synchronism, and it must be tripped. Section 3.4.1 describes how the equal area criterion is used for the stability assessment. If too many big generators trip suddenly, the system can collapse.

Modern microprocessor relays can send the trip command to circuit breakers as fast as in 10 ms, but often it takes a longer time, sometimes even 50 ms, to send a trip command. Adding the relay and circuit breaker operation times together, the fault duration can range from 60 to 100 ms. A typical practice is to plan the grid according to a fixed fault duration of 100 ms. Backup protection systems have adjustable delays since the main protection should trip first. Typical backup protection consists of the second and third zones of distance relays, which send a trip command after around 0.5 and 1 s, respectively.

5.2.2 Protection Reliability

5.2.2.1 Unwanted and Missing Trips

According to IEC Standard 60050-448 [2], reliability of protection is defined as the probability that a protection system can perform a required function under given conditions for a given time interval. In this context, the required function for protection is to operate when required and not operate unnecessarily.

Protection system faults can be divided into two different classes: missing trips and unwanted trips. Unwanted trips can further be divided into spontaneous and unselective trips. A spontaneous trip is a trip without a fault in the system. An unselective trip is an unwanted trip during a fault in the system. If the system is operated according to the $N - 1$ principle, the unwanted spontaneous trips do not cause system problems. The other protection malfunctions can jeopardise the stable operation of the system. A delayed trip is closely related to a missing trip since it can be regarded as a missing instantaneous trip by the main protection relays.

The reliability of protection is divided into two categories based on two fault types: the *security* of protection, defined as the probability of protection from not having an unwanted operation, and the *dependability* of protection, defined as the probability of protection from not having a failure to operate. Security is thus connected with unwanted erroneous trips while dependability focuses on the problem of missing trips after grid faults.

The concept of *hidden failure* in power system protection is defined as follows: "a hidden failure is defined to be a permanent defect that will cause a relay or a relay system to incorrectly and inappropriately remove a circuit element(s) as a direct consequence of another switching event" [3]. Therefore, a hidden failure is an unwanted unselective trip and it remains undetected until some other system event causes the hidden failure to initiate a cascading outage. The concept of hidden failure in the context of cascading outages is discussed in Sect. 2.3.4. Often, if not always, the probability of a hidden failure is a function of line loading.

5.2.2.2 Statistics of Protection Failures

Some statistics show that protection system failures are connected to blackouts. In the United States 49 major disturbances occurred in 1984–1988 and 36 of them had a protection system misoperation as a contributing factor [3, 4]. In most cases, a natural event or a device failure was the initiating event, and the subsequent relay system misoperation contributed to the disturbance. According to more recent data from the North American Reliability Council [5], 10, 40, and 50% of North American disturbances had the misoperation of protection and control systems as a leading cause in 2006, 2007, and 2008, respectively. The figures from the 1980s and 2000s are not directly comparable: a contributing factor is different from a leading cause.

According to Phadke et al. [3], the causes of relay misoperations of electro-mechanical and static relays were application (30%), maintenance (44%), setting (11%), and other (19%). According to a more recent analysis [6], the relay failures observed during grid disturbances in the years 1999–2004 varied according to the voltage level, protected object, country and relay type. The comparison covered the years 1999–2004 and included 9,162 correct trips, 968 unwanted trips, 128 missing trips, and 185 other relay failures (delayed trips or other unspecified cases) in Norway and Finland. A conclusion, common to both countries, was that for microprocessor relays, the share of incorrect settings was 24–29% of the failure causes compared with the share of 5–15% for the conventional (electromechanical and static) relays. Another result was that there are more unwanted than missing trips. The report also states that most unwanted Finnish trips are spontaneous at the highest voltages (400 and 220 kV) and unselective at 110 kV, even though no exact figures are presented.

Johannesson et al. [7] present unreliability statistics of the Swedish protection systems of transmission lines during the years 1976–2002. The percentage of power system disturbances with incorrect protection operations was about 7% and that amount did not increase with the implementation of digital relays. In their statistics, the percentage of unwanted relay operation is 59.3% and the percentage of missing operation is 19.3%. The remaining part comes mainly from failures in reclosing or failed indications..

If a grid is operated according to the $N - 1$ principle, an unselective trip without a power system fault, for example during a relay test, does not cause system problems. If there are one or several unselective trips in addition to the correct trips during a fault, the case can lead to a severe disturbance, since it is an $N - k$ contingency, with $k > 1$.

5.2.3 Generator Trips

Generator trips are caused by the faults of a generator, a turbine, a connection line or a busbar. A trip of a large generator, even without a power system fault, can be severe for the system. Sometimes, a trip of a large generator can be the dimensioning fault of the power system: the contingency that sets the limits for the maximum power transmission in a way Sect. 3.5.1 describes. In order to withstand the generator trip, the system has to have generation and transmission reserves. The internal system of a power station is a complicated system. Many possible failures within a power plant may cause the generator to trip. Transmission system operators usually set some requirements for the generator to tolerate grid faults without tripping, for example that a generator has to withstand a short circuit of certain duration.

Typically, generators have different owners and the statistics of generator trips do not necessarily exist. These statistics would be useful, since generator trips affect the grid reliability. In Europe, according to the EU regulation, the trips of generators bigger than 100 MW must be announced. This information is public, but announcing individual faults does not mean that there are statistics. In North America, NERC publishes a GADS-database of generator trips [8].

A power station is a big and complex system, and different power stations or generators can have different, even individual trip rates. The age of a generator can affect its likelihood to trip; the infant mortality of a new power station and generator may lead to a large trip rate during the first months of the operation. A comparison between the Nordic and North American statistics [9] reveals that there are differences in the unavailability values of generators for different countries and different generator types. Nevertheless, when analysing large power systems, an average failure rate can normally be used.

After a trip of a generator, the system faces a frequency dip. The frequency recovers after the reserves in generation activate and increase their power production. If the reserves are located far away from the tripped generators, the power flows in the system change, which may have an effect on voltage and angle stability. Sometimes also manually activated reserves need to be started in order to recover the frequency or to reduce the power flows within allowed transmission limits.

5.3 Component Failures

This section briefly describes substation component failures and failure rate estimates for them. The focus is on the main components needed for event and fault trees, i.e. the basic context of the PSA approach. Circuit breakers, relays, and telecommunication systems are the main components needed for isolating the faulty parts of the power system.

5.3.1 Failure Rate and Component Unavailability

5.3.1.1 Failure Mode and Effect Analysis

Failure mode and effect analysis, FMEA (introduced in Sect. 4.3.3.2), is used for systematically identifying the different failure modes and their effects, causes and identification of the system components [10]. The data received in this analysis are necessary for selecting the reliability models of components appearing in fault trees. The fault trees can then be built according to the information received in the failure mode and effect analysis and according to the substation structure.

The qualitative failure data of the substation components can be obtained for example by specifications, substation diagrams, and expert interviews. The expert judgments of the specialists in different fields of power system planning, operation and maintenance are important sources for the reliability model. The FMEA data, as well as the structures modelled, are system-specific rather than universal. Different transmission companies may have different substation structures, different protection systems, different maintenance policies and they may have devices manufactured by different companies.

An example of data collected from each device failure

The substation and component location there, component group and class (e.g. switches, circuit breaker), type by the manufacturer, owner, voltage level, commissioning year, manufacturing year, the temperature during the failure, faulted part, the cause of failure, how the failure was detected (e.g. test, substation inspection, alarm, disturbance), the effect of failure, can the component be used when it is failed, the urgency of the repair, class of failure (major, minor, other), repair time, inspection time, identification number for the failure, In addition, there can be specific fields for different component groups. It is useful also to have a field for all kinds of comments and discussions.

5.3.1.2 Failure Rate

Often constant failure rates are used even though typically the failure rates are higher in the beginning and end of the life cycle as Fig. 4.6 presents it. Then, the times between successive failures are exponentially distributed (see Sect. 4.3.3.1). Failure rate estimates can be calculated in two ways. The classical maximum likelihood estimate is

$$\lambda_C = \frac{k}{T_{TOT}}, \tag{5.1}$$

where k stands for the number of failures detected during a certain time period and T_{TOT} is the total number of component-years during that period. A Bayesian failure rate estimate (introduced in Sect. 4.3.3.3), using a non-informative prior distribution [11, 12], can be calculated

$$\lambda_B = \frac{0.5 + k}{T_{TOT}}. \tag{5.2}$$

If many failures were observed, the Bayesian failure rate estimate, Eq. (5.2), approaches the classical estimate presented in Eq. (5.1), which is the number of failures divided by the component-years. If no failures occurred and the number of component-years is small, the Bayesian estimate might provide too large a value. The estimate becomes better as the number of components and component-years increases.

If the user would like to use (constant) unavailability values rather than failure rates for some components, it is possible to calculate the asymptotic unavailability, q_M, caused by the faults detected by an alarm (also discussed in Sect. 4.3.3)

$$q_M = \frac{T_{DT}}{T_{DT} + T_{TF}} = \frac{\lambda \cdot T_{DT}}{\lambda \cdot T_{DT} + 1}. \tag{5.3}$$

where λ is the failure rate ($\lambda = 1/T_{TF}$), T_{DT} is mean downtime, and T_{TF} is mean time between failures [10, p. 370–371]. The downtime T_{DT} consists of the time before the repair starts, time for repairing, and the time before the item is restarted. In practice, $T_{DT} \ll T_{TF}$, and the average unavailability q_M of a monitored component is approximately

$$q_M = \frac{\lambda \cdot T_{DT}}{\lambda \cdot T_{DT} + 1} \approx \lambda \cdot T_{DT}. \tag{5.4}$$

If there is no alarm after the component has failed, but the component is regularly tested, the availability should be calculated differently. The failure remains until the component is tested (or otherwise checked), and the time between the failure and the following test has to be included. According to Lapp [13], if the failure rate is λ and the regular test interval is T_{TI}, the average downtime T_{DT} for a tested component is

$$T_{DT} = \frac{1}{\lambda} \cdot \frac{\lambda \cdot T_{TI} - \left(1 - e^{-\lambda T_{TI}}\right)}{1 - e^{-\lambda T_{TI}}}. \tag{5.5}$$

This is valid if the test and possibly repair durations are negligible, i.e. $T_{DT} \ll T_{TI}$. Substituting the average downtime of a tested component, Eq. (5.5), into Eq. (5.3) leads to the following expression for the average unavailability q_T of a tested component [11, p. 174]

$$q_T = \frac{\lambda T_{TI} - \left(1 - e^{\lambda T_{TI}}\right)}{\lambda T_{TI}} = 1 - \frac{1}{\lambda T_{TI}}\left(1 - e^{-\lambda T_{TI}}\right). \tag{5.6}$$

5.3.2 Circuit Breakers

Circuit breakers display one failure mode that is relevant when considering their operation in isolating the faults and ensuring security. The failure mode is a failure to trip after the breaker has received a trip signal from the protection system. If automatic reclosing functions are included in the reliability model, also the *failure*

Table 5.1 The failure rate estimates for different types of circuit breakers (CB)

Component type	Failure mode	Component-years	Number of failures	Failure rate estimate year^{-1}
Air-blast CB	Trip	377	9	1.7E-02
Minimum oil CB	Trip	508	6	4.9E-03
SF6 CB	Trip	527	2	2.8E-03
Air-blast CB	Reclose	377	7	2.0E-02
Minimum oil CB	Reclose	508	7	1.5E-02
SF6 CB	Reclose	527	5	1.0E-02
A terminal strip of a circuit breaker trip coil or a close coil is disconnected	Trip or reclose	4,236	2	5.9E-04

The failures are either *trip after command* or *reclose after command*. Data received from the Finnish 1993–2002 failure statistics or expert interviews [1, p. 194]

to reclose is relevant. Failure rates should be estimated only for the failures that are included in the analysis; other failures should be excluded.

The device failure database of the Finnish transmission system operator Fingrid Oyj was used in the examples in this chapter. Here, all the similar components (for example microprocessor relays or SF6 circuit breakers) are assumed to have similar failure rate estimates irrespective of the manufacturer. The failure rate estimates for tripping and reclosing of SF6 circuit breakers were 2.8E-03 and 1.0E-02 year^{-1}, respectively. These values were calculated from a 10-year statistics and are presented in Table 5.1. Minimum oil and air-blast circuit breakers had higher failure rates than SF6 circuit breakers. Air-blast circuit breakers need a pneumatic system, too, but the assumption here is that failures of the pneumatic system do not affect trips since air-blast circuit breakers can trip once with their own pressure air; and therefore the breaker failure rate of the circuit breaker alone is relevant to tripping.[1]

If a terminal strip of a trip or close coil of a circuit breaker is disconnected due to a human error, a trip signal from the relays cannot reach the circuit breaker. The trip coils are duplicated and the main relays send their trip signals to redundant coils. Table 5.1 presents the failure rate estimates of circuit breakers, based on the Finnish 10-year failure statistics.

If the terminal trips of a circuit breaker are open, the trip signal does not reach the breaker. This failure was not included in the statistics, but it has occurred, and an estimate, based on expert judgments, is calculated for it, too.

CIGRE, for example, has published circuit breaker failure statistics, where it classifies circuit breaker failures into major, minor and other failures. A major failure is a complete failure, which causes the lack of one or more of its fundamental functions [14, p. 135]. This definition includes both opening and closing

---

[1] Reclosing functions (and the possible following trips) are blocked if the pneumatic system has a leakage. Leakages are quickly observed since they send an alarm.

functions. Therefore, if only one function, for example tripping, is relevant, the failure rate estimate may be too high. Indeed, according to CIGRE survey [14], the failure rate for the major failures of single pressure SF6 circuit breakers is 1.21E-02 year^{-1} at the voltage levels 300–600 kV. This number includes only major failures. If both trip and reclosing failures for SF6 circuit breakers in Table 5.1 were summed, the failure rate estimate would be 1.2E-02 year^{-1}, which is almost equal to the failure rate published by CIGRE. According to Table 5.1, only two out of seven failures were tripping failures.

The failure rates were calculated with the assumption that the failures are detected during tests only. Actually, some faults were detected during a disturbance or when a circuit breaker was needed. Therefore, some failures were detected before the next regular test.

5.3.3 Protection System Failures

For the relays, the only failure that is relevant to the PSA substation reliability model is failing to send a trip signal to circuit breakers when requested to do so. This failure can be caused by an erroneous setting, failures during manufacturing, design, installation, or maintenance. There can be a failure in the contacts between the relay and circuit breaker. Modern relays have self-supervision and send an alarm after some failures. Table 5.2 presents failure rate estimates for line protection relays. Faults that send an alarm due to self-supervision and faults that can be detected only during regular tests are presented separately. Electromechanical relays do not send alarms. The estimates in Table 5.2 present failures at voltage levels 110, 220 and 400 kV and cover a period of 12 years. Components at different voltage levels are treated together since the relays are quite similar regardless of the grid voltage. The data come from the failure database of the Finnish transmission system operator.

The unavailability caused by failures that send an alarm are calculated by using Eq. (5.4), and the unavailability due to other faults with Eq. (5.6). Table 5.3

Table 5.2 The failure rate estimates of line protection relays for failures to send a trip signal to circuit breakers and the corresponding component-years

Component	Component-years	Failures detected by alarm	Failures detected by tests	Failure rate estimate λ_M for faults that send an alarm (year^{-1})	Failure rate estimate λ_T for other faults (year^{-1})
Z microprocessor	3,002	18	21	6.2E-03	7.2E-03
Z static	1,854	13	27	7.3E-03	1.5E-02
Z electromechanical	2,319	–	9	–	4.1E-03
D microprocessor	182	6	3	3.6E-02	1.9E-02

The data covers 400, 220 and 110 kV line distance (Z) and differential (D) relays from the Finnish 1998–2009 failure statistics

Table 5.3 The unavailability values of 400, 220 and 110 kV line distance (Z) and differential (D) relays caused by failures to send a trip signal to circuit breakers

Component	Unavailability q_M (failures that send an alarm)	Unavailability q_T (failures detected by tests)
Z microprocessor	6.2E-05	3.6E-03
Z static	7.3E-05	7.5E-03
Z electromechanical	Not relevant	2.1E-03
D microprocessor	7.1E-04	9.9E-03

The corresponding failure rate estimates are presented in Table 5.2. Data are received from the Finnish 1998–2009 failure statistics

presents unavailability values for relays, caused by the failures detected in different ways and calculated by using the data in Table 5.2. The unavailability is the probability that a relay will not send a trip signal to circuit breakers if a fault occurs and it should do so.

Table 5.3 clearly shows that the unavailability is mainly caused of faults detected by tests. The failures that send alarms are soon repaired and their unavailability is negligible.

Actually, the expressions of q_M and q_T in Eqs. (5.4) and (5.5) correspond to a unique failure mode, in an asymptotic regime. If the same component displays both types of failure modes, the mean downtime due to the combination of failure modes can be estimated. If $T_{DT} \ll T_{TI}$, this total mean downtime is close to the sum of the two mean downtimes. For reliable items, the average unavailability is approximated by the ratio of the mean downtime over the mean time to failure. Therefore, in this case, an acceptable approximation for the total unavailability is the sum of the two unavailability values in Table 5.3.

5.4 Telecommunication

Differential relays always need a telecommunication channel and so do the distance relays, if an instantaneous trip is used along the whole line. Therefore, if a fast trip is essential, the protection systems for lines should be equipped with at least one telecommunication channel. The channel can be an analogue or digital power line carrier (PLC), radio link, optic fibre, or a combination of these. When there is only one telecommunication channel, all the relays at both ends of the line use the same telecommunication channel. This means that, even though the two main relays are redundant, the protection system for faults near the line ends is not completely redundant. Figure 5.1 presents the relay protection system, which has two distance relays at both line ends. Both distance relays at each line end send a trip signal to the line circuit breaker. In Fig. 5.1a, the relays have a common telecommunication channel.

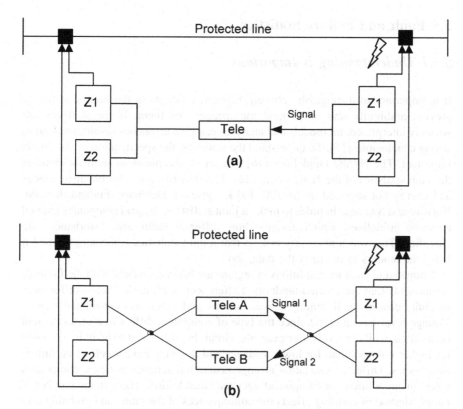

Fig. 5.1 Line protection systems with duplicated distance relays (Z) at both line ends **a** presents a case where the relays have one telecommunication channel and **b** presents a case where the relays have two telecommunication channels. Both relays at each line end send trip signals to the circuit breaker, denoted with a *black square*

Duplicated telecommunication channels increase reliability. Figure 5.1b presents a case where the relays have two telecommunication channels and all the relays can use both channels. In order to be redundant, the two channels always have to have different routes and should have a separate direct current (DC) supply, too. If the availability of one telecommunication channel is 99.7%, for example, the availability of duplicated channels is 99.999%. This is true if the faults are independent. Common components in these two channels significantly reduce the availability of the telecommunication channel.

A power line carrier may be unavailable during line short circuits and earth faults since the faulted phases often fail to transmit the signal. During three-phase faults, all the phases carry the fault current and it is very probable that the telecommunication signal cannot pass through the faulted line. During one- and two-phase faults, it is possible that the healthy phase or phases can carry the signal.

5.5 Fault and Failure Statistics

5.5.1 Understanding is Important

It is important to distinguish between different concepts in the fault statistics to prevent ambiguity and understand the meaning of them. If the statistics are wrongly interpreted, all models and analyses can give erroneous results and lead to wrong conclusions [15]. To understand the statistics, the specification of the data is important. The Nordic Guidelines [16] is a set of specifications used for creating the fault statistics of the Nordic countries. The statistics cover faults, disturbances and energy not supplied in the 100–400 kV grids in Denmark, Finland, Iceland, Norway and Sweden. In order to make a joint statistics, the grid companies created common guidelines, which describe how different faults and disturbances are classified. This was a necessary task so that a joint statistics is meaningful, and it helps the readers to interpret the data, too.

Climate and weather conditions in regions are often correlated with the faults of components that are located outdoors. Differences in climates lead to differences in faults caused by lightning strokes, storms and other weather-related causes. Voltage level has an effect since the type of equipment differs to a certain extent between the voltage levels. For example, circuit breakers tend to be more complex for higher voltages than for lower voltages and this may have an effect on failure frequencies. Different substation arrangements, maintenance policies, component types, manufacturers, or component ages may lead to differences in the number of faults. Method of earthing affects the consequences of the faults and probably also the number of faults.

Some components have different ways to fail; an item can be totally or partly broken, or it can function erroneously. Especially protection devices and switches have different failure modes. When components are part of a dynamic system, the system consequences of the different failures may be different. Component failure statistics seldom give system consequences. Indeed, this would be meaningless if the system consequences varied according to the system state. In a power system, the grid topology and the amount of power flow affect the consequences of faults.

When doing a failure mode and effect analysis, it is important to know the different possibilities of malfunctions, find good statistics, understand them, and use the occurrences of real events as source and interview experts. Power systems are huge and complex systems and no one can know all things in detail.

Such commonly used terms as a fault and a disturbance have different meanings in different contexts. A common way to define a fault is the state of an item characterized by its inability to perform a required function, excluding preventive maintenance or other planned actions, or due to lack of external resources [17]. The North American Reliability Council has a more detailed definition for a (grid) fault: "an event occurring on an electric system such as a short circuit, a broken wire, or an intermittent connection" [18].

There are different definitions also for the term *disturbance*. The Nordic definition for a disturbance is: "outage, forced or unintended disconnection or failed reconnection as a result of faults in the power grid" [19, 20]. NERC [18], on the other hand, gives the following, more detailed, definition for a disturbance:

1. an unplanned event that produces an abnormal system condition,
2. any perturbation to the electric system, or
3. the unexpected change in ACE (area control error) that is caused by the sudden failure of generation or interruption of load.

According to the Nordic definitions, an event is classified to a disturbance only if at least one circuit breaker trips. According to NERC, any perturbation can be a disturbance. Based on the example above it appears that there are different definitions for common concepts. Therefore, it is important to read the specifications of the statistics carefully before using the numbers. Otherwise, there is a risk that the conclusions based on statistics are erroneous and may even lead to incorrect actions.

The quality of data can be dependent on the age of the data. If the structure and material of new equipment are different from the old equipment, old data may not be relevant any more. Again, when using the statistics, one has to understand the contents in order to make correct interpretations.

5.5.2 An Example: The Finnish Line Fault Statistics

This section presents the Finnish 400-kV grid fault statistics with the focus on line faults. As grid faults are the initiating events of the substation reliability model, this section contains a fundamental part of the source data to be applied in reliability modelling.

The organisation of the Nordic transmission system operators, Nordel, published the grid fault and disturbance statistics in 1999–2008. From 2009, ENTSO-E[2] will continue to publish the statistics. An overview of the Nordic grid disturbances is available on the web (for example [20]).

According to Nordel's fault statistics from the year 2008, the average number of faults in the Finnish 400 kV grid was yearly 21.2. This is the 10-year average from 1999 to 2008. This number includes all faults, not just line faults, even though line faults are the most common faults; they represent 73.8% of all faults; 16.8% of all faults were substation faults, most of them belong to category *control systems* [20]. In these statistics, control systems include protection devices [16].

In Finland and many other regions, an earth fault of an overhead line is the most common grid fault and its most common cause is a lightning stroke. The number of

---

[2] In 2009, Nordel's activities were transferred to the new organisation ENTSO-E (European Network of the Transmission System Operators).

faults varies from year to year, mainly but not only, because the weather conditions vary. Figure 5.2 presents the numbers of line faults in the Finnish 400 kV grid during the years 1983–2009. The total line length increased during these years 31%, from 3,479 km in 1983 to 4,570 km in 2009. Figure 5.2 shows that the average (annual) line fault frequencies per line length can have different values depending on the span of years on which they are calculated.

The average annual line fault frequency per line length is

$$f_{\text{LF}} = \frac{\sum_{i=1}^{M} n_i}{\sum_{i=1}^{M} l_i} \tag{5.7}$$

where n_i is the number of line shunt faults during the year i, l_i the line length during the year i, and M the number of years. Thus, f_{LF} is equal to the total number of line faults divided by the total kilometre-years of the lines.

In Finland, the annual line fault frequencies (10-year average) of the 400 kV grid, vary between 0.22 faults per 100 km (years 2000–2009) and 0.35 (years 1994–2003). During the 27-year period presented in Fig. 5.2, there were totally 279 line faults and 103,111 kilometre-years, which give an average 0.27 fault per 100 km per year. During this 27-year period, the share of lightning strokes was 76% and the share of one-phase faults was 75%.

Figure 5.3 presents the division of the faults listed in Table 5.1 according to months. Faults are divided into faults caused by lightning and other faults.

Figure 5.3 shows that most line faults are caused by lightning strokes and occur during the summer months. The distribution of other faults is more constant. If there are seasonal variations in the grid loading and typical power flows, it might be appropriate to use different fault frequencies for different seasons rather than just one constant fault frequency. Using a constant failure rate for all months is good if the loading and power flows are similar regardless of the month.

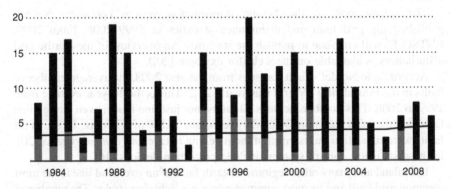

Fig. 5.2 The number of line faults and the corresponding line length in the Finnish 400 kV grid during the years 1983–2009. Each column is for 1 year; the *black part* represents the faults caused by lightning strokes and the *grey* part is for all other faults. The *black line* shows the total length of 400 kV overhead lines in thousands kilometres. The numbers come from the fault statistics of Fingrid Oyj, the Finnish transmission system operator

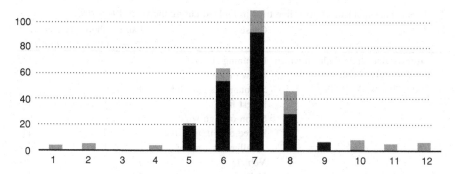

Fig. 5.3 400 kV line trips (224) in Finland during the years 1983–2002 represented with respect to the months. *Dark columns* present faults caused by lightning strokes and *grey columns* present faults caused by other reasons

Most line faults are temporary and high speed or delayed automatic reclosing relays connect the line back into the grid after a short period. This improves the stability and security since the grid becomes rapidly restored after a fault. Table 5.4 presents fault numbers and causes in the Finnish 400 kV main grid during the period 1983–2002. A lightning stroke is the most common fault cause, but there are also several other causes. Some fault causes remain unknown. In addition to line faults, Table 5.4 also presents data with 60 cases where the trip was caused by something that did not happen at the line but at the substation: a substation fault, human error during the relay testing or incorrect operation of a relay. It is not meaningful to include these faults in the calculation of faults per line.

Most line faults were earth faults and short circuits having enough current for distance relays to trip. In addition to them, there were ten high-resistance earth faults, i.e. faults tripped by the sensitive earth fault relays. The distance relays are not capable of tripping these faults. When the fault resistance is high, the fault currents are low and voltages high.

Most faults were successfully cleared by rapid or delayed automatic reclosing relays. The causes of permanent faults were a failure of a tower or part of a tower, a tree, a vehicle that cut the guy wire of a tower, ice on phase wires or dew on earthing wires, a broken tower due to an earth slide due to a nearby dumping place. Once a storm created three successive short circuits before the line was manually disconnected. All these three faults are included in Table 5.4, because they are separate line faults that needed to be tripped.

If the faults are more probable in a certain area, it is relevant to use different fault frequencies in different parts of the grid. This is especially important if the consequences of a fault at a line or a substation are more severe than the consequences of another fault. For example, for a nuclear power station, losing the connection to the grid is critical. Relay misoperations and errors due relay testing present a high proportion of line trips, but if there are no simultaneous faults, the grid can survive.

Table 5.4 Four-hunderd kilovolt line trips in Finland during the years 1983–2002

Fault class	Cause	Earth faults	Short circuits	Other causes	All
An ordinary line shunt fault, in which the main relay (distance or line differential relay) tripped	Lightning stroke	127	38		165
	Snow or ice	3	1		4
	Spontaneous landslide	1			1
	Forest fire		1		1
	Storm or high wind	5	4		9
	A tree or felling a tree	9	2		11
	A failure of a tower	6	1		7
	Vehicle	1	1		2
	Unknown	14			14
	Totally	166	48		214
High resistance earth fault (sensitive earth fault relay tripped)	A tree or felling a tree	7	0		7
	Insulation chain	1			1
	Forest fire	1			1
	Unknown	1			1
	Totally	10	0		10
A fault at the substation or a human error	Explosion of a current or voltage transformer	2	1		3
	Closing a wrong disconnector	1			1
	Temporary earth was connected during the reconnecting the line		1		1
	Circuit breaker series			1	1
Erroneous relay trip	Unwanted spontaneous trip. Most often caused by relay testing			30	30
	Unselective trip during a grid fault			23	23
Other causes	Inrush current of a current transformer			1	1
	All line trips together	179	50	55	284

During these years, there were totally 284 line faults and 72,800 line-kilometres [1, pp. 42–52]

An example of data collected from grid disturbances (partly from Nordel [15])

Fault class (line, busbar, or other fault), identification number, date and time, fault location, faulted component (e.g. line, transformer, series capacitor), fault type (earth fault, short circuit, series fault, other), faulted phases (if relevant), the relays that tripped, how and when the component was reconnected (high speed or delayed automatic reclosing, manual), was the fault primary or a consequent fault, energy not supplied at a delivery point. The causes of a fault should be classified in such a way that it is possible to use the predefined causes unambiguously. Possible choice can be, e.g. lightning, other natural causes (e.g. moisture, ice, natural disasters, pollution, rain, salt, snow, wind, heat, and fire); external influence (e.g. animals and birds, aircrafts, excavation, collision, explosion, tree felling, vehicle, external work, and vandalism); operation and maintenance (e.g. human error of the personnel, an accident during work, maintenance, lack of monitoring, erroneous settings; technical equipment, dimensioning, error in technical documentation, design, temporary constructions, device in general). Also other causes such as installation, production fault, ageing, fault in the neighbouring network, system cause). In addition, it is useful to have a (realtime) field for all kinds of comments and discussions.

References

1. Pottonen L (2005) A method for the probabilistic security analysis of transmission grids. A doctoral dissertation, Helsinki University of Technology, ISBN 951-22-7592-9 http://lib.tkk.fi/Diss/2005/isbn9512275929/. Accessed 29 June 2010
2. IEC 60050-448 (1995) International Electrotechnical Vocabulary. Chapter 448: Power system protection
3. Phadke AG, Horowitz SH, Thorp JS (1995) Anatomy of power system blackouts and preventive strategies by rational supervision and control of protection systems. ORNL/Sub/89-SD630C/1, A rep for the Power Syst Technol Program, Energy Division Oak Ridge Natl Lab
4. Tamronglak S, Horowitz SH, Phadke AG, Thorp JS (1996) Anatomy of power system blackouts: preventive relaying strategies. IEEE Trans Power Deliv 11(2):708–715
5. NERC (2009) System performance indicators: reliability performance gap. N Am Electric Reliab Corp, http://www.nerc.com/page.php?cid=4|37|228. Accessed 10 May 2010
6. Kjølle GH, Gjerde O, Hjartsjø BT, Engen H, Haarla L, Koivisto L, Lindblad P (2005) Protection system faults—a comparative review of fault statistics. The 9th International Conference on Probab Methods Appl to Power Syst, June 11–15, 2006, Stockholm
7. Johannesson T, Roos F, Lindahl S (2002) Reliability of protection systems—operational experience 1976–2002. 8th IEE International Conference on Dev in Power Syst Prot, 5–8 April 2004, pp 303–306
8. NERC (2008) Generating availability data system (GADS), http://www.nerc.com/page.php?cid=4|43. Accessed 29 June 2010
9. Pöyry (2008) Voimalaitosten käytettävyysselvitys. Report for the Finnish Electricity Market Authority, http://www.energiamarkkinavirasto.fi/files/Voimalaitosten_kaytettavyysselvitys.pdf. Accessed 29 June 2010
10. Rausand M, Høyland A (2004) System reliability theory, models, statistical methods, and applications, 2nd edn. Wiley, Hoboken. ISBN 0-471-47133-X
11. Høyland A, Rausand M (1994) System reliability theory. Wiley, New York. ISBN 0-471-59397-4
12. Lee PM (1997) Bayesian statistics: an introduction, 2nd edn. Arnold, a member of the Hodder Headline Group. London. ISBN 0 340 67785 6
13. Lapp SA (1986) Derivation of an exact expression for mean time to repair. IEEE Trans Reliab R-35(3):336–337

14. CIGRE (1994) Final report of the second international enquiry on high voltage circuit-breaker failures and defects in service. Working Group 06 of Study Committee 13, Report 83, http://www.e-cigre.org. Accessed 29 June 2010
15. Evans RA (1999) Stupid statistics. IEEE Trans Reliab 48(2):105
16. Nordel (2008) Nordel's guidelines for the classification of grid disturbances. http://www.entsoe.eu/index.php?id=63. Accessed 29 June 2010
17. IEC 60050-191 (1990) International Electrotechnical Vocabulary. Chapter 191: Dependability and quality of service
18. NERC (2009) Reliability standards for the Bulk Electric Systems of North America. N Am Electric Reliab Corp. http://www.nerc.com/docs/standards/rs/Glossary_2009April20.pdf. Accessed 13 May 2010
19. Statnett and Sintef (2001) Definisjoner knyttet til feil og avbrudd i det elektriske kraftsystemet—Versjon 2. http://www.energy.sintef.no/Prosjekt/KILE/. Accessed 13 May 2010
20. Nordel (2009) Grid disturbance and fault statistics 2008. http://www.entsoe.eu/index.php?id=63. Accessed 13 May 2010

Chapter 6
Substation PSA Model

6.1 Introduction

The chapter describes how to create the substation model, the core of the PSA application, for grid security. The model consists of event and fault trees described in Chap. 4. First, the modelling principles and then the details of the model are presented. The properties of the model are based, to some extent, on the actual Finnish 400 kV transmission grid, which shows that the method is applicable to grids of real size [1]. The example presented to illustrate the method is for line faults.

The steps presented in the chapter and needed in the construction of the substation reliability model are:

- The selection of credible events (faults) for analysis and the estimation of their frequency. Fault frequencies might vary according to the season.
- Defining the substation functionality: the busbar and circuit breaker arrangements that affect the construction of the substation reliability model.
- The selection of grid topologies and power flows included in the analysis.
- Defining the extent of the model, for example the main protection relays, the structure of the protection, fault clearance times, the local and remote backup protection and their functionality, breaker failure protection.
- Defining the principles for event and fault tree construction.
- The construction and analysis of event and fault trees.

The chapter presents first general modelling principles, assumptions and after that, construction of two event trees in detail. The possible failures at the substation (here referred *substation consequences*) are the results of the first event tree analyses. The substation reliability model produces the most probable consequences and their probabilities in the substations after the fault. After the event trees are built, they can be analysed. The results of these analyses are the frequencies of different substation consequences and the local (substation) importance measures. These results can be used for comparing different busbar and relay protection systems but they do not give any indication for power system behaviour. These substation consequences are then the inputs of the dynamic simulations, which give the power system consequences. The dynamic simulations are described in Chap. 7. Briefly, this chapter presents the construction of event trees and the way of finding the substation (or local) consequences. Chapter 7 describes dynamic simulations needed to find the power system consequences.

6.2 Implementation of the PSA Methodology

6.2.1 Event Trees and Their Consequences

Substation modelling here follows the principles of probabilistic safety analysis, PSA, where the accident starts with an initiating event, and the following safety functions and possible consequences are modelled with event and fault trees, described in Chap. 4. This approach is suitable for modelling the power system protection since there is an analogy between the functions at the substations after a grid fault and safety actions of a nuclear power station, where the PSA method has been applied previously.

Figure 6.1 shows a simple event tree, where both substation and power system consequences are added to the end branches. The purpose is to combine the

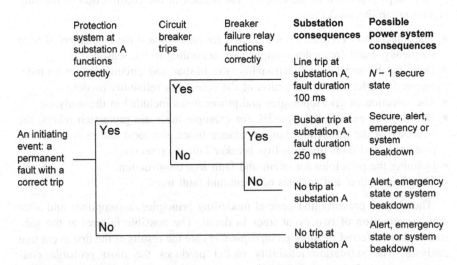

Fig. 6.1 A simplified event tree with an initiating event, success and failure branches of the protection and circuit breakers. The end branches have both substation level consequences (fault duration, tripped components) and a possible choice of system level consequences. Dynamic simulations are needed for system level consequences, which may be different with different power flows

possible failures at the substation and dynamic simulations of the grid together. In this way, the failures at the substation and their power system consequences are connected together, which enables ranking the importance of components at the system level and revealing the chains of events that may lead to a system breakdown. The results of an analytical method, such as PSA, can give in-depth understanding about the system operation and its weaknesses in addition to frequencies or probabilities. Event and fault tree analysis is illustrative and the event trees, when correctly built, give the necessary data for power system dynamic simulations. In addition to this, the results of event tree analyses, i.e. the probability values of the end branches of the event trees, can be used alone for evaluating different substation or protection systems.

In redundant systems, such as meshed transmission grids or double busbar substations, the failures of single components have only minor or no effects on system operation. In consequence, there are not often enough statistical data of the system level failures for the verification of the system reliability. Good operational reliability may have been observed but the system may have hidden weaknesses. If the goal is to find the best options to increase the reliability, the analytical approaches may be helpful. In large systems, there often are statistical data of the failure rates, mean times to repair, or the unavailability values of components. Combining component level data, system structure, and dynamic consequences of faults and failures, an estimation of the system level reliability and the vulnerabilities of the system can be achieved.

Dynamic simulations, described in Chap. 7, are used to determine the power system consequences (the stability and possible violations of voltage, current or frequency limits) after different post-fault failures of the substation components. A power system is a dynamic system, where steady state analyses made with power flow simulations are not sufficient after faults where the protections system has failed and therefore the fault duration is longer, After the dynamic simulations have been executed, their results, *system level consequences*, are added to the event trees as presented in Fig. 6.1.

6.2.2 Modelling Principles for the PSA Model

6.2.2.1 Defining the Scope

Here the system under study is the power system, and the model consists of those components and functions at the substation that are needed to isolate the fault. Therefore, credible power system faults, relay protection systems, circuit breakers, and telecommunication systems are included in the study. When modelling any large, complicated and dynamic systems, the method and model should be correctly focused and consist of only the relevant parts and their connections. One cannot model everything. If the goal is security analysis, the reliability engineer should first define what kind of faults might create a threat to the system security.

One has to select power system topology (intact grid or outages), power flows (defined by the generator injections and load off-takes), season, and other issues that may have effect on the faults and system behaviour after the faults.

Only distance or differential relays are modelled in the event trees presented in this book since these relays are commonly used in the meshed transmission grids. The initiating events here are only those line faults that can be tripped by the distance relays. High resistive earth faults are therefore excluded. They are infrequent and their fault currents are smaller, and often the distance relays are unable to trip these faults.

The event tree branches are constructed in such a way that their analysis of gives the different possible consequences that are necessary for power system analysis. In a power system dynamic analysis, one needs to know the fault duration and the sequence of circuit breaker trips. This principle leads to such a structure that the main protection system and the circuit breakers need to be in separate branches in event trees (also separate fault trees), since the consequence of the failure of the main protection is different from the failure of the stuck circuit breaker. If the main protection systems at one line end fail to send a trip signal, nothing at this substation stops the fault current flowing. If the circuit breaker fails to trip, the breaker failure relay can trip the other circuit breakers connected to the same busbar as the faulted circuit breaker.

Here, the event trees are only for the functions directly connected to the main protection systems and the related circuit breakers. This means that event tree failures are limited to what happens at the line ends of the faulted line, not all the grid locations. 'This assumption is justified because the failures (to trip the fault) are at the substations at the line ends; therefore it is very unlikely to have effects in other locations. This selection limits the consequence analyses, otherwise the event trees and the consequence analyses would explode. Grid dynamics are not included in the event trees but they are calculated separately.

Selected initiating events (here transmission grid line faults), the substation arrangements, and the degree of modelling details define the scope of the model. One has to define the area under study. If the grid under study is a part of a synchronous power system or has strong connections to other systems, one has to decide how to treat the faults at other systems: their consequences should consistently be included or ignored. One option is to first fix the study boundaries and then assume a probability for the external faults, the consequences of which affect the grid under study. With detailed and precise knowledge of the faults outside the study boundaries, detailed models rather than assumptions are possible.

6.2.2.2 Protection and Circuit Breakers

The relay protection system and circuit breakers isolate the faults from the healthy parts of the system. The relays and circuit breakers are located at the substations. The relays are connected on the secondary circuit of the voltage and current transformers, which together with the circuit breakers belong to the primary circuits and often are located outdoors. The relays detect faults, which cause voltages and currents different from normal. The relays then send a trip signal to the circuit breakers, which trip and in this way isolate the faulted part from the power system.

Figure 6.2 presents the line protection system used in the model of this book. The distance relays (Z1 and Z2) are at the substations at the line ends. They receive the current measurement from current transformers (CT) and voltage measurements from voltage transformers (VT). The relays define if there is a fault at the line with the measurements. If they measure a fault, they send trip signals to circuit breakers, which open and isolate the faulty line from the grid. The relays can trip fast after faults at all line locations if there is at least one telecommunication channel between the relays at different line ends.

The reliability models should vary according to the structure of the protection system. The protection of important components is often redundant and has two separate main protection systems. For lines of a meshed grid, distance and differential relays are common since they are able to trip selectively during different grid connections and regardless of the direction of the fault current; a normal case in a meshed grid. Less important objects may have one main and one backup protection system. Different remote and local backup protection systems exist. One has to decide how many details, if any, of the backup protection is included in the model. Automatic reclosing operations after line trips may or may not be included, depending on the details of interest.

The PSA model for a substation consists of event and fault trees. Event trees model the (successive) events of a sequence while the fault trees give the success and failure probabilities of the event tree branches. In principle, the combination of event and fault trees can be different. Having a complicated event tree combined with simple fault trees is one option. A detailed fault tree with simple event threes is the other possibility. For example, two redundant relays can be in the same fault tree in one event tree branch or they can be in two different fault trees and in two

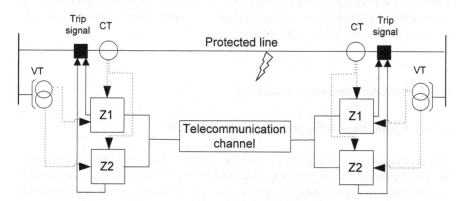

Fig. 6.2 A typical line protection system. *Z1* and *Z2* distance relays, *CT* current transformer, *VT* voltage transformer. When the relays detect a line fault, they send trip signals to circuit breaker, which are represented by black squares

event tree branches. One way to make the choice is defining how many different failures at the substations (the end branches of event trees) are sufficient for a comprehensive dynamic stability analysis. Therefore, different substation consequences should be listed first since they define the substation model structure. In addition to this constraint, the aim is to make the event trees (the chains of events) as simple as possible and to ensure that the analysis of event trees gives the necessary data for power system analysis.

The event tree branches are in this book constructed in such a way that their analysis gives the different possible consequences that are necessary for power system analysis. In a power system dynamic analysis, one needs to know the fault duration and the sequence of circuit breaker trips.

6.2.2.3 Instrument Transformers

The voltage transformer is common for both distance relays at the same line end, as Fig. 6.2 shows, but voltage transformers are not modelled. Distance relays can be connected to different secondary circuits of voltage and current transformers. Then a failure in a measurement circuit affects on one relay only. The distance relays can trip during a fault even if the voltage measurement is lacking, therefore the failure of a voltage transformer does not necessarily prevent the trip. The voltage transformer's supervision systems of both relays could operate during a voltage transformer failure and prevent a trip. However, these occurrences send an alarm and are detected, after which the line is disconnected. It may be relevant to think that both the probability of these occurrences is small and the duration of the failures is short; thus they can be ignored.

The current transformer's primary coil is a common component for the two main protection relays. In Finland, several failures of current transformers were observed. They were all explosions and created a substation fault. Many of these faults were tripped by a busbar protection relay. If there was a line fault during the explosion, there would be two simultaneous grid shunt faults, which is a different situation than is modelled in this study.

6.2.3 Assumptions and Accuracy

6.2.3.1 Connections between Substations and Grid

The assumption here is that the successes or failures of the substation components are not dependent on the power system states, power flows and dynamic stability or instability. In other words, the operations at the substation do not depend on the power flow or dynamic states or transitions of the grid. The assumption is that the probabilities of the failures of the substation components, what it comes to missing trips, are not dependent on the power system state or dynamics. Hence, possible failures at substations and their probability are dependent only on the equipment properties and the substation and protection system structure. The failure modes received in the failure mode and effect analyses for the substation components did not reveal significant[1] failures that would depend on the grid state. There were only few power line carriers and none of them was the only telecommunication channel.

Similar event trees are valid for all those 1-, 2- and 3-phase faults that can be tripped with distance or differential relays, i.e. faults with fault resistance smaller than some 20 Ω. The assumption is that the functioning of substation components after faults is not dependent on the fault current phase and magnitude. The substation model only includes the main protection systems and breaker failure protection, no other backup protection systems are included. These assumptions are relevant since the high resistance earth faults, which require different protections systems are not among the initiating events considered in this book. They are rare and their effect on stability is small compared to arc faults.

The grid topology changes according to the connection or tripping of circuit breakers. Occasionally, components are disconnected from the grid for maintenance. This may affect the substation model; depending on the topology, different components need to be tripped after faults. The consequences of a similar fault followed by the same incorrect protection operation may have different system level consequences during different power flows and different grid topologies. Therefore, different substation models and dynamics simulations might be needed for different cases.

6.2.3.2   Exceptional Arrangements

In addition to normal and typical lines and substations, which are modelled with similar principles and similar fault and event trees, the grid may have unusual or rare substation arrangements. When modelling these, one should think about the purpose of the model. If the purpose is the system security (for example the stability after faults), one should model in detail those parts that affect the security but one can ignore issues that do not have effects on security. An example of this is a special 3-branch line, in which two line ends have normal bays with busbar arrangements and protection systems, but the third branch is feeding a local load via a 400/110 kV transformer without a normal bay. This branch has only a circuit breaker and instrument transformers but no busbar. The scheme of this kind of substation is presented in Fig. 6.3.

---

[1] The unavailability of power line carrier telecommunication is dependent on the state of the line, failed or not failed. During 3-phase faults, all the phases carry the fault current and it is very probable that the telecommunication signal cannot pass the faulted line. In this case, the constant unavailability of the telecommunication is 1. This failure has a correlation with the state of the grid, therefore constant unavailability is not a correct way of modelling although often used.

Fig. 6.3 A schematic figure
of a 3-branch line

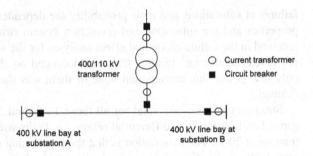

A fault at the 3-branch line can be isolated if circuit breakers at three locations trip. Therefore, the protection system is different from a normal line with two line ends. In spite of this, the event trees of 3-branch lines are constructed in such a way that the protection operations of the transformer branch are ignored. This branch has protection systems, for example two distance relays. If the trip of this branch were to fail, the fault current infeed from the 110 kV grid via the transformer would be small due the high impedance of the transformer. This small current does not have any effect on the grid dynamic stability and therefore it is not necessary to consider it when analysing the stability. Therefore, the line can be treated in a similar way as 2-branch lines. If there was a generator behind the 400/110 kV transformer with a high fault current infeed, it would not be relevant to ignore the branch.

6.2.3.3 Focus to Relevant Issues

Different simplifications prevent the event trees from being extremely complicated. The assumption that if a circuit breaker is stuck, all the phases are stuck, simplifies event trees. This is a conservative assumption; it would be more probable that one phase of the circuit breaker would be stuck rather than all. However, if all the different possibilities of each circuit breaker (1, 2, and 3 stuck phases) were taken into account, the number of event tree function events and branches would increase dramatically. However, the extra information received with this method would be secondary. If the system withstands a failure where three phases are struck, it certainly withstands failures with one or two phases stuck, too. There probably are some locations where the power system consequences would be different with different numbers of stuck phases. With this modelling simplification, the extra information caused by these differences is lost. Each time there is this kind of conflict one has to decide if the extra information received with a more detailed model would be significant and relevant.

Another way to keep the event trees simple is the concept of *fatal failure*, after which no other branches are added; the failure branch after a fatal failure always is the end branch. Figure 6.1 has two branches of this kind, labelled *No trip at substation A*. Such fatal failures are here those where the fault current continues to flow at one line end, i.e., either the main protection relays fail to send a trip signal to circuit breakers or a breaker failure relay fails to trip correct circuit breakers after a circuit breaker stuck. From the power system point of view, there is no need to know if the trip succeeds at the other end if it has failed at one end. The failure is fatal enough and it is quite insignificant what would occur at the other line end. Additionally, a simultaneous 'no-trip' failure at both line ends would have the probability that is the product of the probabilities of both line ends, a very small number.[2]

In this book, in order to avoid ambiguity, the concept of backup protection is only for those relays that operate if the main protection relays fail to operate. Therefore, the instantaneous settings of two distance relays at a substation, both protecting the same line, belong to the main protection system. The other zones, those that send delayed trip signals (second and third zones of the same relays) belong to backup protection, even though all the zones exist in the same relay. Breaker failure protection also belongs to a backup system that isolates a busbar if a circuit breaker fails to trip.

6.3   Event Trees

6.3.1   Components of the Substation Reliability Model

6.3.1.1   Components Modelled

There are three main components, the reliability of which is important when studying the power system functioning after a fault and which are modelled. The components are

- the relay protection system, which includes the relays and the secondary circuits of the protection,
- the circuit breakers and their connection to the relays via wires, and
- the telecommunication system between the relays.

After a fault in a power system, the protection system detects the fault and sends a trip signal to circuit breaker(s). Then the circuit breakers trip in order to isolate the fault, after which the power flow can continue in the healthy parts of the power system. The breaker failure protection and the backup protection systems enter into action if the main protection or the circuit breakers do not function properly. The event trees are created for substation events, taking into account three main parts listed above.

---

[2] The local or remote backup protection is not in included in the event trees and therefore, in the model, the fault current does not stop if the main protection or circuit breakers fail. If the operation of backup protection is of interest, one can include them in the model.

6.3.1.2 Initiating Events

A reasonable estimate for the initiating events (definitions, frequencies) is needed for good results. Common and severe credible faults should be used. Usually, if not always, overhead line faults are more common than substation faults, since lines are extended to large geographical areas and are situated outdoors. Both qualitative (causes of faults) and quantitative (fault numbers) analyses are useful for defining those credible grid faults that are included. Statistics that cover long rather than short periods provide better estimates. Figure 5.1 clearly shows both the stochastic character of line faults and how a selection of a short time period may lead to too high or too low fault frequency estimates. The possible differences of fault frequencies in different seasons or at different lines are useful, if there are differences in fault frequencies according to season and if this data is available.

The substation model presented in this book corresponds to line faults. There are two reasons for this choice. They are the most probable grid faults [1]. In many grid locations, line faults are dimensioning faults too, which gives another reason to look at them before other faults.

Each event tree needs an input, an initiating event, which in this book is a line shunt fault. Here the assumption is that the average (annual) line fault frequency is constant per unit line length. The estimate for an initiating event (a line fault frequency) is therefore the product of the line fault frequency and the length of a line section. The long-term annual average frequency of line faults f_{LF} is calculated with Eq. 5.8. Letting the length of a line section of event tree k be l_k, the frequency of the initiating event of the event tree k is

$$f_k = l_k \cdot f_{LF}. \tag{6.1}$$

6.3.1.3 Line Protection with Distance Relays

This chapter briefly describes the line protection of distance relays, necessary for understanding the substation model. A schematic presentation of line protection with distance relays is in Fig. 6.2.

The line protection realised with distance relays is based on *zones*, which define which faults locations belong to the relay and how fast the relay sends a trip signal. Each zone has a reach and trips the faults located at its zone. Figure 6.4 presents the typical zones of a distance relay. Zone 1 covers some 80% of the protected line. The relay sends an instantaneous (not delayed by the settings) trip signal if it measures a fault at its zone 1. Zone 1 cannot cover 100% of the protected line, since there are errors in the voltage and current measurements and with a setting of 100%, the relay would trip faults at the remote end substation or behind that. Zone 2 usually covers the protected line and a section of the adjacent lines. The reach can be for example 120% of the protected line. Zone 2 has a delay (for example 0.4 s) and sends a trip signal after the delay. The delay means that if the fault is at the adjacent line, the relays protecting that line trip first. Zone 3 reaches longer

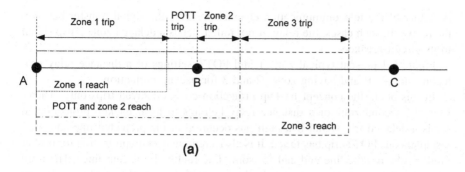

(a)

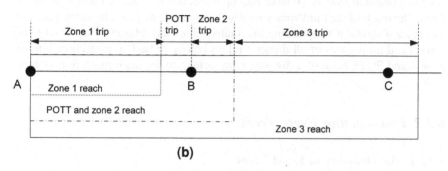

(b)

Fig. 6.4 Typical zones of a distance relay at substation A. A distance relay sends a zone 1 trip signal if it measures a fault in zone 1. It sends a trip signal to circuit breakers if it measures a fault in POTT zone and, in addition, receives a signal form the relay at substation B. When a relay measures a fault in zone 2 or in zone 3, it sends a delayed trip signal after a delay of 0.4 or 1 s, respectively. **a** Presents a case where the backup zone 3 does not completely cover the adjacent line B–C. **b** Presents a case where the backup zone 3 covers the adjacent line B–C

than zone 2 and has a longer delay, too. A typical delay is 1 s. Zone 3 should not reach longer than the shortest adjacent line in order to prevent unselective trips. Sometimes, but not always, zone 3 totally covers the adjacent line, and is a backup relay for faults along that line.

A telecommunication signal between the relays at different line ends enables an instantaneous line trip for all the fault locations. Line faults near the remote line end (seen by the relay) do not belong to the first zone and therefore an instantaneous trip requires telecommunication between the relays at different line ends. A *permissive transfer trip scheme* (POTT),[3] for example, functions so that if a relay measures a fault in its POTT zone and it receives a signal from the remote line end, it trips. This trip is slightly slower that the first zone trip due to a minor delay of the telecommunication. A typical reach of POTT zone is similar to the reach of zone 2.

---

[3] Permissive overreach of a distance protection uses telecommunication and with overreach setting in each relay, and where a signal is transmitted between the relays when a fault is at the overreach protection zone. The receipt of the signal at the receiving end permits sending the trip signal to circuit breakers. (IEC 60050-448).

A failure of the telecommunication channel prevents the signal transfer between the relays. In such cases, the relay, if not failed, sends a delayed zone 2 trip signal to its circuit breakers.

Figure 6.4 present typical zone 1 and POTT settings of a distance relay protecting line A–B and having zones 2 and 3 for backup protection.

In this book, the concept backup protection covers delayed trips only. Therefore, the second zone of a distance relay belongs to local backup protection; it sends a delayed trip signal if a fault has occurred at the protected line and if an instantaneous POTT trip has failed. It is also a remote backup protection for busbar faults at the remote line end and for some line faults of adjacent lines. The third zone of a distance relay is a remote backup protection for the line faults of adjacent lines. The reach of the third zone varies and it cannot always totally cover adjacent line since it should not trip during line faults behind any adjacent line. Briefly, the second and third zones of all distance relays belong to backup protection, but the zone 1 and POTT zone of a distance relay belong to the main protection system.

6.3.2 Constructing Event Trees

6.3.2.1 The Structure of Event Trees

In this substation model, the event trees (presented in detail in Chap. 4) are for analysing the main operations after a grid fault: protection systems and circuit breakers. Each branch of an event tree has one success and one failure path. With the success or failure branches of the main components (listed in Sect. 6.3.1.1), it is possible to create all the post-fault events. Fault trees are used as the inputs of the branches of event trees. The failure probability of each branch is calculated with fault trees; therefore, a fault tree top gate is the input of each event tree branch and gives the success and failure probabilities. The start of an event tree also needs an input. Here the frequency of a line fault is used as input. Initiating event frequency multiplied with the probability of an end branch of event trees gives the frequency of the failure.

Here, the basic structure of event trees is such that the function events of the main protection operations are put before the function events of the circuit breakers, which corresponds with the real-time sequence. Automatic reclosing operations are excluded, since they are not important for grid security immediately after faults; what matters for dynamic stability are correct trips after faults. The events of both line ends are put in the same event tree, one after the other. In the real world, the order of the trip signals of the main protection at different line ends is arbitrary, and the same yields for the circuit breakers trips. This difference between the model and reality does not matter. Since, basically, the event tree is a logical diagram simultaneous (or almost simultaneous) events can be put in an arbitrary order.

The event tree branches are here constructed in such a way that their analysis gives the different possible consequences that are necessary for power system dynamic analysis. In a power system dynamic analysis, one needs to know the fault location, fault duration, and the sequence of the trips of circuit breakers. This principle leads to such an event tree structure that the protection systems and the circuit breakers are modelled in separate branches since the consequence of protection failures and the consequences of a stuck circuit breaker are different.

6.3.2.2 Protection in Event Trees

Faults at different locations of a line require different treatment in the reliability model, because the distance protection responds differently according to the fault location. The distance relays can trip only the faults in the middle section of the line instantaneously without a telecommunication channel between the relays at two line ends. This means that the protection system acts in different ways according to fault location along the line.

The reach of the instantaneous tripping zone of a distance relay (zone 1) is about 80% of the line length. Therefore, 20% of the line at the remote end does not belong to the zone 1 of the relay. An instantaneous trip of the remote end section is possible if the relay receives a telecommunication signal from the relay at the remote line end as Fig. 6.2 and Sect. 6.3.1.3 describe it. A common practice is to use so called permissive overreach transfer trip (POTT) scheme, where a relay trips if it measures a fault and receives a telecommunication signal. Therefore, without telecommunication, the distance relays at both line ends, can instantaneously trip faults located in the middle section (section 1 in Fig. 6.5). The remaining 40% is divided into two sections as shown in Fig. 6.5. Relays at substation B trip fast faults in section 2 if they receive a telecommunication signal from the relays at substation A and vice versa. One event tree for each line section in Fig. 6.5 is needed.

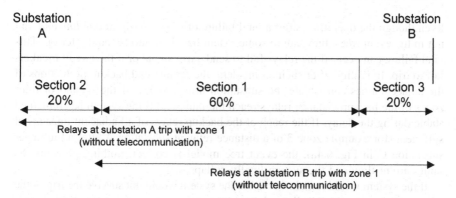

Fig. 6.5 Distance protection and its zones along a line. Zone 1 of a distance relay reaches 80% of the protected line, therefore faults at the remote line ends require a telecommunication signal between the relays in order to provide an instantaneous trip. The faults at the middle section (60%) can be tripped at both substations without telecommunication

Here, the probability of the relays failing to send a trip signal of different zones due to a relay failure is similar since all the relay components (except of the relay setting and a piece of software) are the same. The probability of a trip signal by the permissive overreach transfer trip scheme (POTT, sometimes also POP, permissive overreach protection) depends on the telecommunication channel, too. The telecommunication channel is modelled as a separate component. If a distance relay is healthy but fails to receive a telecommunication signal due to a failure of the telecommunication channel, the relay can send zone 2 trip signals, delayed with the zone 2 settings when there is a fault at the remote end of the line. Therefore, the event trees representing the sections of the line ends have an additional branch for the delayed zone 2 trip signal. Figure 6.4 presents typical distance relay settings at substation A protecting line A–B and having zones 2 and 3 for backup protection for line A–C.

The operation of a distance relay is about the same for 1-, 2- and 3-phase faults. In a reliability analysis, it is a relevant assumption that the relays succeed or fail regardless of the number of faulted phases unless better information exists. In this example, only 3-phase faults are considered, though.

The breaker failure relay is a relay that operates if a circuit breaker is failed and does not trip after it has received a trip signal. It measures the current of the circuit breaker that has received a trip signal. If the current does not stop in a given time (for example 200 ms), the breaker failure relay trips all the circuit breakers connected to the same busbar as the faulted one and sends a trip signal to the distance relays at the remote end substation of the faulted line bay. Figure 3.9 illustrates the operation of a breaker failure relay.

If the local protection systems at one line end fail to send a trip signal, nothing at *this substation* stops the fault current flowing. This is considered as a fatal failure.

6.3.2.3 The Concept of a Fatal Failure

Even though the operations after a fatal failure of type '*no trip at one line end*' are not in the event trees, they can to some extent be taken into account later on. This fatal failure can occur if the relays fail to send a trip signal or if the circuit breakers fail to trip. In reality, after such an incident, the remote end backup protections of the adjacent lines can isolate the substation after a delay, if the second or third zones of the backup distance relays reach the fault, and if the system has remained stable during the delay. If the reach of the backup relays of all adjacent lines is not sufficient (for example zone 3 of a distance relay at substation A after a fault near substation C in Fig. 6.4a), the event tree model is correct: nothing prevents the fault current flow and the system finally collapses.

If the system would remain stable but the system would not survive the trip of the whole substation with all its lines, the power system consequence would be the same: a system breakdown. Only if the system does not lose its stability during the long fault duration and after the system has lost the whole substation after the trip by the backup relays, the system may survive. This could be for example a fault at line B–C in

Fig. 6.4b, after which a relay at substation B would fail, but zone 3 of a relay at A would trip. Remote backup protection systems of this kind are not modelled in event trees, though. Modelling these would increase the branches of event trees drastically but the probabilities of the additional branches would be very small.

6.3.2.4 Different Event Trees

The details of the event tree depend both on the substation arrangements at the line ends and on the fault location. The substation arrangement has an effect on the number of circuit breakers at the line ends. Each circuit breaker at a line end leads to two event tree branches, one for the circuit breaker and the other for the breaker failure relay. Figures 3.8 and 3.9 illustrate busbar arrangements.

There are three different lines when classified according to the number of circuit breakers. Different fault locations along the line lead to different event trees; the difference being the possible need for a telecommunication. The event trees with the smallest number of branches are for cases where the fault in the middle of the line (no telecommunication needed) and there is only one circuit breaker at each line end. Three different event trees for each line lead to nine different event tree constructions. Table 6.1 presents different choices.

In Table 5.1, event trees 1, 2 and 3 are for the lines where both line ends have two circuit breaker arrangement, similar to arrangements in Fig. 3.9 b, d. The event trees have the biggest number of different substation consequences and event tree branches. Event trees 4, 5 and 6 are for lines, where one line end has two circuit breakers and the other line end has one circuit breaker. These fault trees (for example 4a and 4b) have some mutual symmetry, and they have an equal number of consequences. The event trees have the failures of two circuit breakers at one line end only, thus the number of consequences is smaller than in event trees 1, 2 and 3. Event trees 7, 8 and 9 are the simplest since they have the smallest number of circuit breakers and breaker failure relays.

Table 6.1 Different event trees for line fault analyses

Event tree number	Line type	Fault location
ET 1	Both line ends have two circuit breakers	Middle of the line
ET 2	Both line ends have two circuit breakers	Near LE1
ET 3	Both line ends have two circuit breakers	Near LE2
ET 4a	LE1 has two and LE2 has one circuit breakers	Middle of the line
ET 5a	LE1 has two and LE2 has one circuit breakers	Near LE1
ET 6a	LE1 has two and LE2 has one circuit breakers	Near LE2
ET 4b	LE1 has one and LE2 has two circuit breakers	Middle of the line
ET 5b	LE1 has one and LE2 has two circuit breakers	Near LE1
ET 6b	LE1 has one and LE2 has two circuit breakers	Near LE2
ET 7	Both line ends have one circuit breaker	Middle of the line
ET 8	Both line ends have one circuit breaker	Near LE1
ET 9	Both line ends have one circuit breaker	Near LE2

LE1 is line end 1 and LE2 line end 2

6.4 Event Trees in Detail

6.4.1 Fault in the Middle of the Line

In the following, one event tree is presented in detail. This event tree is developed
for a case with single circuit breakers at both line ends. The fault location is in the
middle of the line and the distance relays at both line ends measure the fault as
being on zone 1; therefore, the model does not contain any telecommunication
channels. The event tree, ET7, is presented in Fig. 6.6.

The branches of the event tree are the following, starting from the initiating
event:

1. The main protection system of LE1 (line end 1) sends an instantaneous trip
   signal to the circuit breaker. If this fails, no other operations are checked; this is
   a fatal failure.

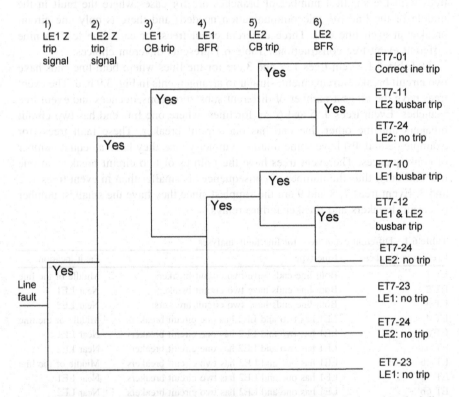

Fig. 6.6 An event tree for modelling the operations after a line fault in a case where both line
ends have one circuit breaker. The fault location is in the middle of the line. This is ET7 in
Table 5.1. *LE1* and *LE2* line ends 1 and 2, respectively. *Z* distance relay, *CB* circuit breaker, and
BFR breaker failure relay. The *numbers of the end branches of event trees* are codes for different
consequences

2. The protection system of LE2 (line end 2) sends an instantaneous trip signal to the circuit breaker. The failure branch of this is the end branch since this failure is regarded as fatal.
3. The circuit breaker at LE1 trips. The failure of this branch leads to the branch for breaker failure protection.
4. The breaker failure protection of LE1 trips the relevant circuit breakers, in a similar way as Fig. 3.9 c presents. If this fails, there is no trip at this substation and the failure branch is the end branch.
5. The circuit breaker at LE2 trips. The failure of this branch leads to breaker failure protection.
6. The breaker failure protection of LE2 trips the relevant circuit breakers. If this fails, there is no trip at this substation and the failure branch is the end branch.

A description of each end branch and the consequences connected with the branches are presented in Table 6.2. A consequence of an event tree is a kind of label attached to each end branch. The consequences in this stage are substation

Table 6.2 Description of the end branches of event tree ET7, presented in Fig. 6.6 and in Table 6.1

The end branch of the event tree in Fig. 6.6	The consequences at the substations
7-01	LE1 and LE2 (line ends 1 and 2) succeed in tripping the faulted line. The fault duration is 100 ms after which the line trips at both ends. This is an $N - 1$ secure case and the grid remains stable. Automatic reclosing actions are not considered
7-10	At LE1, the protection system functions correctly, the circuit breaker fails to trip but the breaker failure protection succeeds in tripping the busbar after the delay defined by the breaker failure relay and the circuit breaker operation time (typically 250 ms). At LE2, the protection and circuit breakers succeed in tripping the fault in 100 ms
7-11	At LE1, the protection and circuit breakers succeed in tripping the fault after 100 ms. At LE2, the circuit breaker fails to trip faulted line, but the breaker failure protection succeeds in tripping the busbar after the delay defined by the breaker failure relay and the circuit breaker operation time (typically 250 ms)
7-12	One busbar is tripped at both substations by the breaker failure relays. The relays at both line ends send trip signals, the circuit breakers at both line ends fail to trip but both breaker failure protection systems succeed in tripping the relevant busbars after the delay defined by the breaker failure relay and the circuit breaker operation time (typically 250 ms)
7-23	The relays at LE1 fail to send trip signals to the circuit breakers, or the circuit breakers fail to trip after they have received a trip signal. Nothing at LE1 stops the fault current
7-24	The relays at LE2 fail to send trip signals to the circuit breakers, or the circuit breakers fail to trip after they have received a trip signal. Nothing at LE2 stops the fault current

consequences. It is worth noting that the same consequence can be caused due to different failure sequences. The substation consequences are independent of the load flow case because they depend only on the successes and failures at the substation components after the fault. The substation consequences include both the fault durations and the tripped components. The consequences are numbered. The numbers, descriptions and explanations of consequences are presented in Table 6.2 of event tree 7 (presented in Fig. 6.6).

6.4.2 Faults at Line Ends

For line faults near the line ends, there are three protection branches instead of the two presented in Fig. 6.6. These are:

- The protection system at the line end near the fault sends an instantaneous (zone 1) trip signal.
- The protection system at the other line end sends an instantaneous permissive overreach transfer trip (POTT) signal.
- The protection system at the other line end sends a delayed (zone 2) trip signal.

Figure 6.7 presents an event tree (ET8), where the fault is near LE1 (line end 1). Compared with Fig. 6.6, an extra branch, there is an extra branch, a delayed zone 2 trip signal, located after the failure of the instantaneous trip signal. Because of this, there are more end branches compared with event tree ET7 in Fig. 6.6. After the success branch of the POTT trip signal and the success branch of the zone 2 trip signal, a similar set of circuit breaker and breaker failure protection branches are needed. This is a common property of event trees: adding a branch often doubles the number of end branches.

Consequences of event tree ET8, presented in Fig. 6.7, are listed in Table 6.3.

If the line has more circuit breakers than event trees ET7 and ET8 in Figs. 6.6 and 6.7, extra branches are added to the event trees. Each circuit breaker needs a branch for a trip and another branch for the breaker failure relay.

6.5 Fault Trees

6.5.1 General Principles

Usually, the basic events in the fault trees are for a single component, such as a certain relay or a certain circuit breaker. These basic events depend only on the component itself. However, some basic events are for common components, the failure of which affects several fault trees. Such a basic event corresponds to, for example, the substation batteries and the direct current systems, which feed the protection relays, and the trip and close coils of circuit breakers. Another example

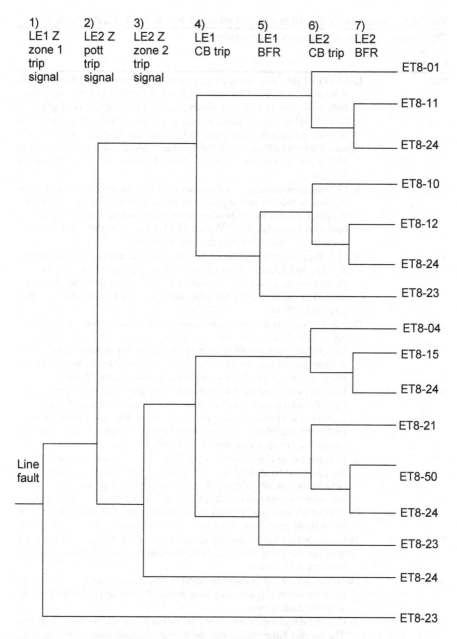

Fig. 6.7 An event tree for modelling the operations after a line fault in a case where both line ends have one circuit breaker and the fault location is in near line end 1, ET8 in Table 5.1. *LE1* and *LE2* line ends 1 and 2, respectively. *Z* distance relay, *CB* circuit breaker, and *BFR* breaker failure relay. The *numbers of the end branches of event trees are codes* for different consequences

Table 6.3 Description of the end branches of event tree ET8 in Fig. 6.7 and in Table 6.1

The event tree end branch in Fig. 6.7	The consequences at the substations
8-01	LE1 and LE2 (line ends 1 and 2) succeed in tripping the faulted line correctly without a delay. The fault duration is 100 ms after which the line trips at both ends. This is an $N-1$ secure case, and the grid remains stable
8-04	At LE1, the relays trip the line correctly in 100 ms. At LE2, the relays fail to send an instantaneous trip signal (POTT), but they send a delayed zone trip signal after which the circuit breaker trips correctly. The fault duration at LE2 is the delay of zone 2 and circuit breaker operation time, for example 450 ms
8-10	At LE1, the protection system functions correctly, the circuit breaker fails to trip but the breaker failure protection succeeds in tripping the busbar after the delay defined by the breaker failure relay and the circuit breaker operation time (typically 250 ms). At LE2, the protection and circuit breakers succeed in tripping the fault in 100 ms
8-11	At LE1, the protection and circuit breakers succeed in tripping the fault after 100 ms. At LE2, the circuit breaker fails to trip faulted line, but the breaker failure protection succeeds in tripping the busbar after the delay defined by the breaker failure relay and the circuit breaker operation time (typically 250 ms)
8-12	One busbar is tripped at both substations by the breaker failure relays. The relays at both line ends send trip signals, the circuit breakers at both line ends fail to trip but both breaker failure protection systems succeed in tripping the relevant busbars after the delay defined by the breaker failure relays and the circuit breaker operation time (typically 250 ms)
8-15	At LE1, the relays and circuit breakers trip the line correctly in 100 ms. At LE2, the relays send a delayed zone 2 trip signal, the circuit breaker fails to trip but the breaker failure relay trips the busbar. The fault duration at LE2 is the sum of zone 2 delay, the delay of the breaker failure relay, and circuit breaker operation time, for example 600 ms
8-21	At LE1, the circuit breakers fail to trip after it has received a trip signal from the relay; the breaker failure relay trips the circuit breaker. The fault duration is typically 250 ms. At LE2, the relays fail to send an instantaneous trip signal (POTT), but they send a delayed zone trip signal after which the circuit breaker trips correctly. The fault duration at LE2 is the delay of zone 2 and circuit breaker operation time, for example 450 ms
8-23	The relays at LE1 fail to send trip signals to the circuit breakers, or the circuit breakers fail to trip after they have received a trip signal. Nothing at LE1 stops the fault current
8-24	The relays at LE2 fail to send trip signals to the circuit breakers, or the circuit breakers fail to trip after they have received a trip signal. Nothing at LE2 stops the fault current
8-50	At LE1, the relays send a correct trip signal but the circuit breaker fails to trip. The breaker failure relay trips the busbar; the fault duration is typically 250 ms. At LE2, the relays send a delayed trip signal after which the circuit breaker fails, but the breaker failure relay trips the busbar. The fault duration at LE2 is the sum of zone 2 delay, breaker failure relay delay and circuit breaker operation time, for example 600 ms

The consequences are listed in the same order as they appear in Fig. 6.7

of such a basic event is the substation pneumatic air system, which produces compressed air for the air-blast circuit breakers.

Some devices send an alarm when there is a failure. Here it is assumed that this alarm is always sent successfully to the control centre, i.e., the alarm acts with 100% reliability. Some devices that send an alarm are not modelled at all. An example of such a device is a rectifier of a direct current (DC) battery. After the DC rectifier has sent an alarm and during the repair, the batteries can keep most of their stored energy. Thus, rectifier failures are not included in the model.

All the relays of the same class (microprocessor, static, or electromechanical) have the same failure mode and effect analysis (FMEA) data irrespective of the manufacturer. The same principle is applied for air-blast, minimum oil and SF6 circuit breakers, too. This is justified if we think that the construction and type rather than the manufacturer of equipment are dominant in failure modes and frequencies. If the situation is different, the model should be built differently.

It is assumed that, if a relay succeeds in sending a zone 1 trip signal to the circuit breaker(s), it also sends, if needed, a zone 2 trip signal, and a signal to the breaker failure relay. A fault tree of breaker failure includes both the relay and the circuit breakers connected to the same busbar as the faulted circuit breaker. This is a simplification, but prevents the model from becoming too complicated. However, in the model, only the circuit breakers of the lines and generators are included since these bays feed large fault currents. The circuit breakers from 110 kV grid via the 400/100 kV transformers are excluded, since the fault current in-feed from 110 kV grid to 400 kV is insignificant compared with the fault current at 400 kV grid and would not risk the stability.

A modelling detail of SF6 circuit breakers is worth mentioning as an example. After commissioning, the SF6 circuit breakers are provided with a blocking system that prevents the trip if the SF6 gas density in the circuit breaker is too low. If the circuit breaker trips with a low SF6 gas pressure, the circuit breaker can be broken. In Finland, the blocking operates only during the guarantee period since it is removed afterwards. The reason for this is that it is better to let one circuit breaker break when tripping rather than let the power system collapse due to continuous fault current flow. Therefore, most SF6 circuit breakers are not provided with this blocking system and blocking system is not modelled in the fault trees.

The parameter *mean time to repair* (see Sect. 4.3.3.1), needed when calculating the average unavailability, is a tricky issue for the grid stability analysis. Here, it has two different interpretations. It is the mean active repair time for those components, whose failure does not cause the changes in the grid topology and connection. The repair duration of each failure is reported in the device failure database and it is possible to calculate the average times for repair time. These kinds of components are the ones that are redundant, for example relays. For some single components, the value of 15 min is used. This is due to the Nordic requirement that the operation be restored back to secure within 15 min following faults that change the grid state from secure to alert. If a single component, like a circuit breaker, is found broken, the faulted circuit breaker is disconnected from the grid and possibly some alterations in the grid loading are made as well. This means that the faulted case duration is only 15 min, since the grid topology, and maybe the grid loading, are changed after that time. This is an approximation of the reality with the purpose to use relevant repair times in the model.

6.5.2 Fault Trees Give Event Tree Probabilities

The branches of the event trees need an input in order to calculate the branch probabilities. Fault tree top gates are the inputs of event tree branches. Fault trees are logical diagrams (already presented in Sect. 4.3.2) and they consist of basic events and gates. The top gate of a fault tree is the input of a branch in an event tree and gives the probability for the success and failure of that branch. Different fault trees made for different protection and trip functions are:

- Two main protection relays at one line end fail to send a zone 1 (or zone 2) trip signal to the trip coils of a circuit breaker. Event tree branches 1 and 2 in Fig. 6.6 and branches 1 and 3 in Fig. 6.7 are for this event.
- Two main protection relays at one line end fail to send an instantaneous permissive overreach trip signal (POTT) to the trip coils of a circuit breaker. Event tree branch 2 in Fig. 6.7 is for this event.
- A circuit breaker fails to trip after it has received a trip signal to its trip coils. Event tree branches 3 and 5 in Fig. 6.6 and 4 and 6 in Fig. 6.7 are for this event.
- The breaker failure protection fails. The fault tree includes both the relays and the relevant circuit breakers, which reduces the possible failure combinations (and loses information about the details), but helps to keep the event trees simple. The failure of this fault tree occurs if one or more circuit breakers remain closed or if the breaker failure relay fails to operate. Event tree branches 4 and 6 in Fig. 6.6 and 5 and 7 in Fig. 6.7 correspond to this event.

One fault tree is in Fig. 6.8. The top gate of this fault tree is the failure of two main protection relays to send a trip signal, either zone 1 or zone 2, to circuit breakers. The common cause failures in Fig. 6.8 represent such faults that can prevent all the components in a substation or a bay to function. For example, a fire that would damage all the cables between a bay and control building can be a common cause failure for a bay. The miniature circuit breakers of voltage transformers or direct current batteries prevent the distance relays from sending a trip signal; thus they are included. The faults of the relay itself are in the basic events Z1-relay and Z2-relay of Fig. 6.8.

6.5.3 Constructing Fault Trees

Usually, there is different data available of different devices and the components are different, too. Therefore, the basic event types can be classified in different

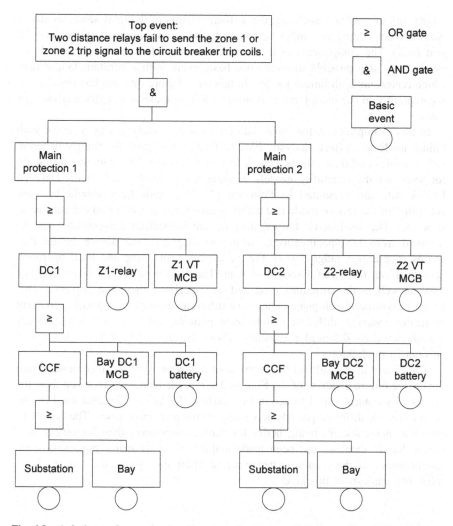

Fig. 6.8 A fault tree for a redundant line protection system. *Z* distance relay, *MCB* miniature circuit breaker, *VT* voltage transformer, *DC* direct current, and *CCF* common cause failure

groups. Some components are modelled simply with constant unavailability. Monitored components, i.e. those that send an alarm after they have failed, can be modelled with failure rates and repair times. The components that are regularly tested and do not send an alarm, can be modelled with failure rates, test intervals and repair times.

The basic event with constant unavailability has the unavailability value as its only parameter. It is the simplest model available. In this example, this model is used for telecommunication channels and for static and microprocessor relays. Static and microprocessor relays have two kinds of failures. A failure in the power supply unit of a relay is such that the self-supervision of the relay sends an alarm. An erroneous setting or configuration can be detected during a test (or during a grid fault). The components of this kind can be modelled in two separate basic events. It is also possible to create one basic event with a constant failure rate, which represents both failure modes. In this way, failure rate and test interval are not explicitly in the model and it is not possible to make sensitivity analyses for them.

In this example case, the input data for fault tree analysis was received with failure mode and effect analysis, FMEA. FMEA was used to identify different failure modes and their effects, causes and identification. This data was necessary for selecting the reliability models of components in the fault trees. Detailed FMEA data are presented in Pottonen [1]. The fault trees should be built according to the failure mode and effect analysis and according to the substation structure. The qualitative failure data of the substation components can be received from the specifications, substation diagrams and device failure databases. The expert judgments of the maintenance, planning and local operation personnel can be a useful source of data. The failure mode and effect analysis data, as well as the structures modelled, are specific rather than universal. Different transmission companies may have different substation structures, different protection systems, different maintenance policies and they may have devices manufactured by different companies. These differences have to be taken into account in modelling.

The quantitative results of the example of this book were received mostly from the device failure database of the Finnish TSO. Some data were received from the supervisory control and data acquisition system (SCADA). The data used in this book cover the different periods depending on the respective cases. The quality of data was not constant, being better for some components than for others. The device failure database has been made mainly for maintenance purposes. Some interpretation of the data was necessary in order to classify the failures for a reliability analysis of this kind.

6.6 Common Cause Failures

In addition to separate components the failure of which can lead to missing trips, it is important to include common cause failures if such exist. The model presented in this book does not contain common cause failures other than the substation and the bay. These common cause failures are basic events in Fig. 6.8. The common cause failure *bay* consists of the cable ditch and all the cables between the control room and bay. There is a risk that something might damage all the cables in that ditch and prevent the transmission of the signals. This failure would prevent all operations of a bay: protection, circuit breaker trip, communication between the bay and control room.

The basic event for the whole substation is in all fault trees of that substation. This basic event models a fire, for example, or a mechanical failure of the substation or in the building. All the secondary systems of the substation (relays, telecommunication devices, part of the cables connecting the bays to the secondary system) are in the substation building.

A common cause failure due to maintenance or testing is one question that needs consideration. If the terminal strips of both trip coils of a circuit breaker are disconnected, the circuit breaker cannot trip, even though it would be undamaged. If all the terminal strips of all the circuit breakers of a bay were disconnected due to an error in maintenance or testing, this would be a common cause failure. This could happen if the maintenance of all circuit breakers of a substation was carried out simultaneously.

Another tricky question is the possible common cause failure of the two microprocessor relays by the same manufacturer protecting the same line. If two redundant relays come from the same manufacturer, a common cause failure is possible. Different relays by the same manufacturer may have the same software parts even though they were of different types. It is possible that, in this case, there is a possibility for a common cause failure. A common cause failure for two microprocessor relays by the same manufacturer would be the best way to model this, but finding a correct value for this would require more information than is usually available for the relay user. A miniature circuit breaker (MCB) of the redundant distance relays is also an example of a common cause failure is the device is common for both relays.

6.7 Analysis of Event Trees

The event trees built here are analysed according to the principles presented in Chap. 4. The results are frequencies (or probabilities[4]) of the end branches of the event trees, the minimal cut sets of different failures, and different importance analyses. The results of different event trees are separate and give information about the chains of events described in the event trees. It is possible to use the substation model as such and analyse substation failures. Pottonen et al. [2, 3] presents some examples of such comparisons.

No power system consequences can be received, nor any system level analyses made until the power system consequences of different end branches of the event trees are known. Chapter 7 describes the dynamic simulations and how their results are connected to the substation model.

---

[4] If the initiating event is modelled with a probability rather than a frequency, the end branch result is a probability, too.

References

1. Pottonen L (2005) A method for the probabilistic security analysis of transmission grids. A doctoral dissertation, Helsinki University of Technology, 951-22-7591-0, 951-22-7592-9 http://lib.tkk.fi/Diss/2005/isbn9512275929/. Accessed 29 June 2010
2. Pottonen L, Pulkkinen U, Koskinen M (2004) A method for evaluating the reliability of protection. In: 8th IEE Int Conf on Dev in Power Syst Prot, 5–8 April 2004, pp 299–302
3. Pottonen L, Pulkkinen U, Koskinen M (2004) The effect of relay protection on the substation reliability. In: PSAM7—ESREL04, international conference on probabilistic safety assessment and management, Berlin, Germany 14–18 June 2004, 1002-1007

Chapter 7
Dynamic Consequence Analysis

7.1 Introduction

This chapter presents the combination of the reliability model and dynamic simulations of the power system. This chapter describes the simulations that are made according to the results of the substation model presented in Chap. 6. The chapter illustrates how to find out the power system consequences of the substation failures after line faults.

The chapter describes how the simulations are performed, how the results of the grid dynamic simulations of the chains of events (defined by the end branches of the event trees) are made. The end branches of event trees define different chains of events, most of them with extended fault durations and additional trips caused by backup protection systems. It also presents how to classify the results of dynamic simulations by using different power system states. In this way, the simulations reveal which chains of events lead to a system breakdown and which lead to other system states.

An essential part of the chapter is to show how to combine the substation PSA model and dynamic simulations. The substation model, used for analysing the protection and tripping operations at the substation after a line fault, produces the substation consequences, which are used as inputs to power system dynamic simulation. After the simulations, the results (power system states) are added as new consequences to the event trees, which are analysed again. The results of the analysis with power system states are the probability of the selected power system states and importance values of substation components.

Combining the power system consequences of the failure sequences and the substation PSA model enables the connection of the substation component failures and the power system breakdown. This means that it is possible to make meaningful importance analyses of the failures of the substation components; for example to find out which are the components that, if they fail, create the biggest risk to the system stability. After the combination, the calculation of the frequency

L. Haarla et al., *Transmission Grid Security*, Power Systems,
DOI: 10.1007/978-0-85729-145-5_7, © Springer-Verlag London Limited 2011

of post-fault power system states and different importance measures at grid level are possible.

In this chapter, the whole process for some end branches of an event tree is presented. The purpose is to illustrate how the modelling and analysis were done. The example presented here is, again, the security analysis of the Finnish 400 kV transmission grid after line faults. The results received with the model are presented in Chap. 8.

7.2 Process Description

Figure 6.1 presented a simplified event tree with both local (substation) and global (power system) consequences. Figure 7.1 presents a block diagram of how to first get the substation consequences, then find out the power system consequences, and finally combine these two to the event trees of the substation reliability model (Figs. 6.6 and 6.7).

Only those substation consequences are of interest where the post-fault system state is not known in advance. Depending on the scope, all or some failure branches can be selected. The simulations, the grid model and the definition of the power system states after the fault are presented in detail in Sect. 7.3.

Figures 6.7 and 6.8 present event trees with substation consequences, which are the same regardless of the power flow or grid connection. The frequency of different consequences varies from line to line and is dependent on the components used at the line substations and on the line length, which affects the fault frequency. The power system consequences, on the other hand, are different for different lines since the location of the line affects the stability and other power system phenomena. Therefore, similar substation failures at different line ends lead to different power system consequences. The power system consequences are also a function of the load flow and grid topology. A substation consequence can, during one power flow, lead to a critical safe operation and during another power flow to a system breakdown. Therefore dynamic simulations are needed for defining the power system consequences at different grid connections and load flows.

7.3 Simulations

7.3.1 Assumptions

Before performing the simulations, a set of assumptions of different issues such as the performance of the power system during transients and modelling of backup protection systems in the simulations are needed. Here, for example, an assumption is that only transmission system malfunctions are considered and power plants

Fig. 7.1 The block diagram of the power system reliability analysis after line faults

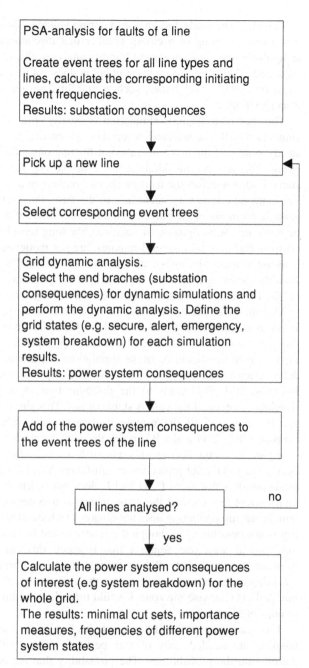

perform perfectly and the generators trip only if they lose angle stability, not due to other reasons, like voltage dips or fluctuations. This means that the modelling should be according to the requirements set by the grid codes.

In reality, the loads, too, may trip during voltage dips or voltage fluctuations. It is not only opening of a circuit breaker that disconnects the loads. In dynamic simulations that are for transient stability analyses, this load property is not usually modelled. Therefore, dynamics simulation does not give energy not supplied, they focus on system level issues, such as stability, not on local issues, such as load disconnection.

The start of a dynamic simulation is usually similar: at first there is pre-fault simulation with the selected power flow to ensure that simulation starts from a steady state and there are no inherent fluctuations power flows due to the modelling. This has certain duration, for example one-second, after which the fault starts. Following that, the trips of circuit breakers in a correct order are simulated. The fault durations vary according to the case, but every case is simulated long enough, normally in the range of 20 s in order to see if the post-fault situation will be stable or not. In dynamic simulations, the long-term issues, for example turbine controls that may take several minutes, are not modelled. Line heating, sag (time constant of tens of minutes), are not modelled either. The purpose is to find out the stability issues right after the faults and let the power system operators to act later on (on the level of minutes) and take care of long-term operative actions.

One should define how to treat the fatal errors of event trees, such as *no trip at a substation*, numbered to 23 and 24 in Figs. 6.6 and 6.7. Figure 7.2 helps to illustrate how to simulate a case where circuit breaker CB2 trips correctly but no trip occurs at substation A. In the simulations, CB2 trips after 100 ms. After a 1-s delay, circuit breakers CB3, CB4 and CB5 trip and isolate substation A totally from the grid. This leads to the disconnection of several lines, and dynamic simulations reveal if the case is stable or not. This simulation should be done if all the 3rd zones of all the remote backup distance relays (connected to circuit breakers CB3, CB4 and CB5) reach the fault location.

Figure 7.2 helps also to illustrate such a case, where the circuit breaker CB2 operates correctly but protection at substation A or CB1 fail. If the remote backup protection at substations C, D and E does not reach the fault, nothing stops the fault current flowing and the consequence is a system breakdown. Zone 3 of the remote backup protection does not always reach the fault location; in this case, the trip is not possible even though the relays would be healthy. If the faulted line is long and at least one adjacent line is short, this can be the case. Figure 7.2 illustrates this possibility. Relays connected to circuit breaker CB3 in Fig. 7.2 have a maximum reach, which is less than the sum the length of line A–C and 80% of line A–E. (Otherwise the zone 3 would reach beyond line A–E, which would lead to non-selective trips.)

The assumption is that if the remote backup protection systems and their circuit breakers are needed, they operate perfectly. The assumption is relevant when considering the probabilities. The probability that remote backup protection and protection at the remote line end would fail simultaneously which would be insignificantly small.

As already mentioned in Sect. 6.2.3.2, if a circuit breaker fails to trip after it received a trip signal the assumption is that all the phases remain closed. In the

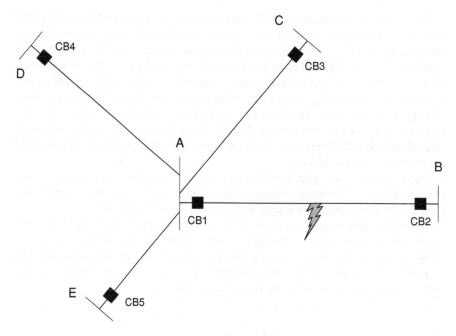

Fig. 7.2 A fault on line A–B after which circuit breaker CB 2 trips in 100 ms. If circuit breaker CB1 fails to trip, the fault current flows via lines C–A, D–A and E–A. If the 3rd zone of the distance relays connected to circuit breakers CB3, CB4 and CB5 is sufficient, these circuit breakers trip after the delay of the 3rd zone, typically 1 s. Otherwise nothing trips the fault

simulations, this means that the 3-phase fault continues until the breaker failure relay trips the busbar, or if this fails, the relays at the remote end trip the whole substation. Other options are possible and can be simulated. If it was probable, that only one phase of the circuit breaker would be stuck, it might be better to simulate a 1-phase fault after two phases are opened.

To simplify the simulations and to reduce the number of simulated cases it is possible to simulate only 3-phase faults with zero fault impedance. If the grid is effectively earthed, the power system effects a 1- and 2-phase shunt faults can be quite similar to those of 3-phase faults.

7.3.2 Simulation Cases

All those consequences of event trees should be simulated where the power system state could not be directly concluded. There is no need to simulate the ordinary $N - 1$ faults (the success branches of event trees) if the grid is planned and operated according to the $N - 1$ principle. The number of substation consequences to be analysed with dynamic simulations depends on the size of the grid and the scope of the analysis.

The dynamic consequences of the component failures connected to grid faults may vary according to the grid topology and power flow. If the power flows vary, the dynamic analyses should be done for relevant power flows. If different fault levels are connected to certain power flow patterns, for example according to the season, the frequencies of the initiating events, and event tree end branches vary, too, and so do the probabilities of a system breakdown and other power system consequences. In such circumstances, it is possible that different substation components create the main risk for security during different power flow situations since the number and frequency of the end branches with a system breakdown as a consequence may be different.

The topology of the grid and the grid state are needed for power system simulation. Different substation arrangements should be included in the model. Some typical substation arrangements are presented in Figs. 3.8 and 3.9.

After the different chains of events after faults (from event trees) are known and the load flow case is selected, the faults can be simulated. The effects of the faults on the power system are known only after the dynamic simulations are made. The severity of the faults can be ranked in different ways. Here the dynamic consequences are classified to different power system states.

7.3.3 Fault Locations and Durations

The line sections that require different event trees are presented in Fig. 6.3. The fault location in a dynamic simulation has to be fixed. One way is to locate faults in the middle of each section. The line sections that require telecommunication are 20% of the line length and situated at the line ends, which leads to a fault location of 10% from the line end. The fault location for the line section where the distance relays do not need a telecommunication channel is in the middle of the line. The fault location for this section is in the middle of the line. The fault locations for simulations are presented in Fig. 7.3.

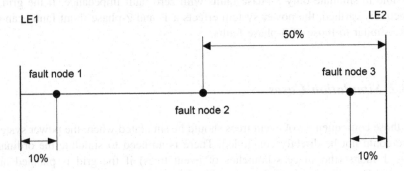

Fig. 7.3 The selected fault locations of a transmission line for the dynamic simulations. Fault node 1 is for event trees 2, 5 and 8, node 2 for event trees 1, 4 and 7 and node 3 for event trees 3, 6 and 9 in Table 6.1. LE1 is line end 1 and LE2 is line end 2

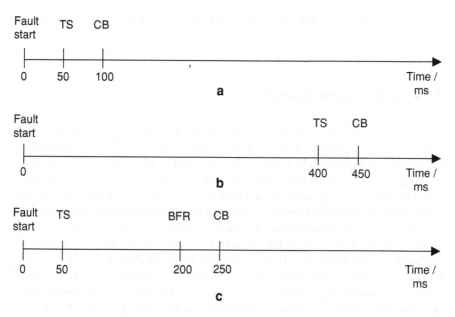

Fig. 7.4 Different fault durations **a** presents two possibilities: a normal trip with the fault duration of 100 ms and a delayed trip with the fault duration of 450 ms. A delayed trip occurs if the fault is at the remote line end and the telecommunication signal fails **b** presents a case where the circuit breaker fails to trip after an instantaneous trip signal from the relays, after which the breaker failure relay (BFR) trips the busbar. **c** The fault duration in this case is 250 ms

In the real world, there are some variations in fault durations, since different relays start and succeed in sending the trip signal in different times, and so, too, do the circuit breakers. There also are variations in relay operations due to different fault locations. In the simulations, however, the fault duration times of the dynamic simulations are fixed and they depend on the relay settings and circuit breaker characteristics. Figure 7.4 presents typical choices for fault durations.

Figure 7.4a presents an instantaneous line trip with the duration of 100 ms, where the relay operations take 50 ms and the circuit breaker trip takes another 50 ms. A trip of this kind can be either a zone 1 trip or permissive overreach transfer trip scheme of the distance relays or a trip by the line differential relays. Figure 7.4b presents a delayed zone 2 trip of the distance relays. The relay sends the trip signal after 400 ms from the fault start; the circuit breaker operation takes 50 ms. The fault duration of this delayed trip is a consequence of the telecommunication failure when the fault is at node 1 and after the telecommunication failure, the relay at LE2 trips with zone 2 settings. In these cases, the line trips and no additional trips occur.

When the breaker failure protection relay trips the busbar after a circuit breaker failure, the fault duration is assumed to be 250 ms. The relays send a trip signal to circuit breakers and to the breaker failure relay after 50 ms. If the fault current has not stopped after 200 ms, the breaker failure relay opens all the circuit breakers connected to the same busbar as the stuck breaker. This case is presented in Fig. 7.4c.

7.3.4 A Simulation Example

Here a description of an event tree consequence ET7–10 (Fig. 6.7) is described. After 1 s, a 3-phase line fault occurs in the middle of the line. The first section of the fault lasts 100 ms after which line is disconnected at LE2 (line end 2) since the protection and circuit breakers there succeed. Now the fault current continues to flow from the grid to the fault via LE1 (line end 1). This fault section lasts 150 ms, which is 250 ms from the fault start. Then, at LE1, the busbar connected to the same busbar as the stuck circuit breaker is disconnected by the busbar protection. Since all the lines are connected to the substation via one circuit breaker, all the lines connected to that busbar are tripped, too. The post-fault state is simulated 20 s in order to see if the power system remains stable or not and if there are violations of voltages or line loadings. Figures 7.5, 7.6 and 7.7 present typical dynamic simulation results. Figure 7.5 presents simulations after a line fault that is correctly tripped. The duration of the fault is 100 ms after which the line is disconnected. The amplitude of the oscillations is not very high and the oscillations decay fast.

Figure 7.6 presents a case where the fault is tripped correctly at one line end but at the other line end the circuit breaker is stuck. After 250 ms, the circuit breakers isolate the busbar where the faulty circuit breaker is. The duration of the voltage dip is longer near the stuck circuit breaker than near the other end of the line. The amplitude of the oscillation is larger and the oscillations last longer than in the previous case. This disturbance is more severe than a normal $N - 1$ contingency but in this case the system remains stable.

Figure 7.7 presents a simulation where the power system loses its angle stability due to the long fault duration. Generators lose their synchronism and the system collapses via oscillations.

7.3.5 Looking at the Simulation Results

7.3.5.1 Principles

The results of the dynamic simulations can be classified taking into account the stability, the voltage violations and the thermal limits, and possibly also the reach the remote backup protection. It is a good practice to rank all the simulation results to different classes in a systematic way and according to specific predefined criteria. The simulation results sometimes require interpretations. For example, if the oscillations after the fault do not decay enough, a certain damping ratio, the limit of a stable case, should be defined.

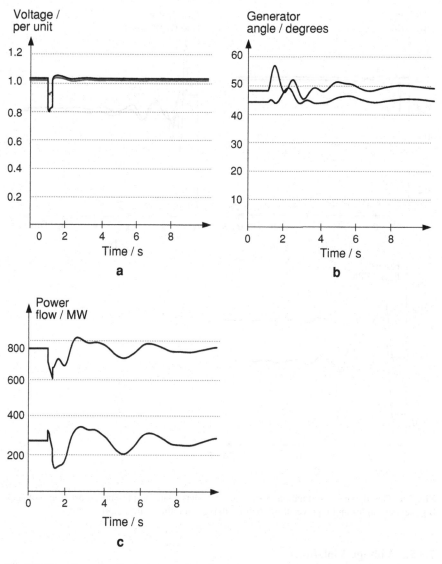

Fig. 7.5 The dynamic simulation results of a secure state **a** presents substation voltages **b** generator angles and **c** power flows before, during and after a line short circuit

7.3.5.2 Stability

Angle stability is one aspect in classification: two options are possible: stable and unstable. When the angle stability of the power system is lost in the simulation time (within 20 s after the fault), the result is a major disturbance. The phenomenon is so fast that the control centre personnel could do nothing to prevent a system break-down. Voltage or frequency stability problems should also be checked.

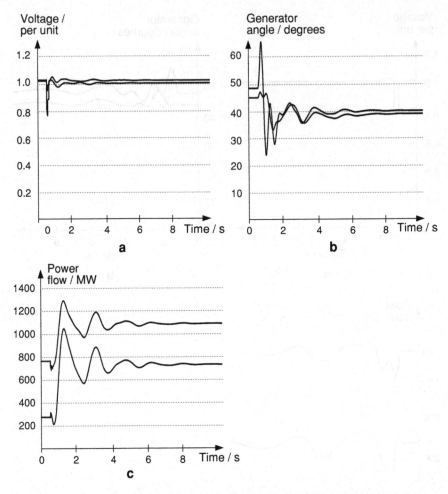

Fig. 7.6 The dynamic simulation results of an alert state **a** presents substation voltages **b** generator angles and **c** power flows before, during and after a line short circuit

7.3.5.3 Voltage Violations

Voltage violations should be checked since it is possible that the voltages can be beyond the limits without any stability problems. The upper voltage limit is defined by the insulation co-ordination and is typically 420 kV for a 400 kV grid. The lower voltage limit should also be specified. A possible choice is 370 kV due to possible voltage instability and voltage quality. The simulated voltages depend on the load models used. Constant power loads produce the lowest voltages since the load does not reduce with decreasing voltages. The simulation model should have correct models and values for the reactive power limits of the generators and generator excitation controls.

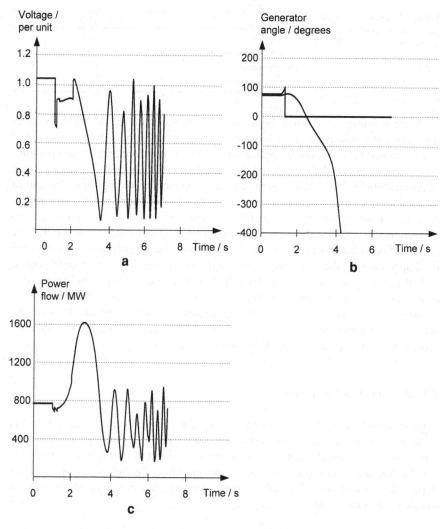

Fig. 7.7 The dynamic simulation results of a system breakdown. **a** Presents substation voltages **b** unstable generator angles and **c** presents power flows before, during and after the fault. The simulation ends up in a system breakdown since no stable post-fault state is reached

7.3.5.4 Thermal Limits

Thermal limits depend on the ambient temperature and on the wind. Typically, short lines and series compensated lines may have thermal limits defining their transmission capacity but stability sets the limits for long lines. The temperature at which the limits are checked should be defined. If the power flow used in the simulations is typical of winter, a winter temperate, for example −10°C, can be used.

Table 7.1 Typical thermal loading values for devices located outdoors at different ambient temperatures

Component	+30°	+10°	−10°
Conductors	1	1.25	1.5
Current transformers	1	1.15	1.3
Disconnectors	1	1	1
Series capacitors	1	1	1

When defining the thermal limits for each line, all the components in series (line, current transformer, disconnectors and series capacitors) should be checked and the lowest limit should be applied for the line. If there are parallel components, for example two parallel current transformers in a two-circuit breaker arrangement (presented in Fig. 3.8b), the current via one component would optimally be half of the total current. It is possible to assume that the current in one parallel component in a substation is not more than 60% of the line total current. There are no measurements made on this subject; it might be too optimistic to assume that the current of one branch would be exactly half since a very small difference between the impedances of the joints may cause different power flows for parallel routes.

If the rated power or current of the line conductors, current transformers, disconnectors and series capacitors for a +30° ambient temperature are known, it is possible to calculate the corresponding values for other ambient temperature values, for example for −10°. Table 7.1 presents typical thermal limits for lines and current transformers situated outdoors.

7.3.5.5 Energy not Supplied

If the dynamic model is such that the energy not supplied can be calculated; this could part of the classification. When the stability is the limiting factor for the transmission, it is usually not possible to get the amount of energy not supplied via simulations since the model does not correctly reflect the behaviour of the loads during the oscillations. Typically, energy not supplied is a measure used with steady state simulations.

7.3.6 Classification of the Simulation Results

7.3.6.1 Power System States

The power system states can be mapped as secure, alert, emergency and system breakdown, as described already in Sect. 2.4. In the secure state, a power system can withstand single contingencies because the $N - 1$ criterion is applied in planning. The alert state seems to be a secure state but there are no longer sufficient margins to withstand an outage due to a disturbance and therefore a loss of a component would result in a current or voltage violation or even a system breakdown. In an emergency state, no load is curtailed, but operating constraints have been violated. When classifying the consequences of the dynamic simulations, it is possible to use these or some other classes, for example an additional case, where extra generators or high voltage direct current (HVDC) links are tripped due to exceptionally long fault duration, if this result is of interest. The classification of power system consequences may vary according to system properties. The classification of this book adapts the Nordic Grid Code [1].

In this book, the results of the simulated cases are classified to the following classes: secure, critical safe operation, limits exceeded and system breakdown. In the example presented in this book, the simulation results after line faults are classified to these four power system states according to the rules presented in Sects 7.3.6.2–7.3.6.6.

7.3.6.2 Secure

Secure means stable operation according to dynamic simulations, i.e., no extra generators or high voltage direct current links were tripped, and no thermal or voltage violations occur. The fault tripped in 100 ms and the therefore the line can be reclosed by high speed or delayed automatic reclosing relays or manually.[1]

7.3.6.3 Critical Safe Operation

The critical safe operation of a power system is a state which is stable, no voltage or thermal ratings are violated, but there were some failures after the fault. Therefore, several lines or a busbar are tripped or the fault duration was extended. A busbar trip and trips of extra generators due to long fault duration belong to this state as well, if the voltages and thermal ratings are not violated. This state is a critical safe operation because the grid might not withstand another fault. This resembles an alert state of a power system.

In reality, many simulation results listed here as critical safe operation could be secure. It is possible that, after some line trips at a certain grid loading, the grid would withstand another fault, which would mean that it is secure. In order to avoid laborious dynamic simulations, the simplification of the model, to label some consequences without simulations, is acceptable. Here, the main interest is to get an overview of the power system reliability and not study all the possible contingencies in detail. Getting an overall idea of the consequences of different contingencies requires many simulations, even with this degree of accuracy.

---

[1] If the line is tripped by line protection relays, automatic or manual reclosing is performed. Automatic reclosings are not performed after a breaker failure relay or remote backup protection has tripped the line. Therefore, these cases are alert rather than secure.

The consequences of *no trip at a substation* (23 and 24 in Figs. 6.7 and 6.8 and in Tables 6.2 and 6.3), where the trip fails at one line end, were classified as critical safe operation if the requirements were fulfilled, and if the reach of the remote backup protection was sufficient to trip the substation, if needed.

An example of a critical safe operation is the tripping of a busbar by the breaker failure relay at a substation with a double busbar arrangement, where each line end has two circuit breakers. A substation of this kind is in Figs. 3.9b and 3.9d. The power flow does not change after the fault is cleared since all the lines and substations are in use after the trip of busbar. The only difference compared to the secure state before the fault is the reduction of security: the faulted substation would lose all the lines connected to it if a busbar fault were to occur before the disconnected busbar is reconnected.

7.3.6.4 Limits Exceeded—Emergency State

A stable state where voltages or thermal ratings or both are outside the acceptable limits is here called a state with limits exceeded. A trip of the faulted line only cannot lead to a state like this due to the $N - 1$ principle. A busbar trip, a substation trip, and trips of extra generators could be classified to limits exceeded—emergency state if the voltages or thermal ratings are violated.

The consequences of end branches *no trip at a substation* (consequences 23 and 24 in Figs. 6.7 and 6.8 and in Tables 6.2 and 6.3), belong to a state *limits exceeded*, if there are violations in the voltages or current limits and the reach of the remote backup protection relays is sufficient (as described in Sect. 7.3.1).

7.3.6.5 System Breakdown

The system breakdown can be due to different causes. An unstable case in dynamic simulations was one reason for a system breakdown. Another possibility for a simulation case to be classified as a system breakdown was such that zone 3 of the remote backup protection did not reach the fault location after the failure *no trip at the substation* (consequences 23 and 24 in Figs. 6.7 and 6.8 and in Tables 6.2 and 6.3). In this case, it did not matter if the dynamic simulation result was unstable or not. If there was no trip at the faulted line end, and additionally, if the remote backup protection did not reach to the fault, nothing else would isolate the fault and system would have a breakdown.

7.3.6.6 Additional States

An extra class, for example a *partial system breakdown*, can be used, too. The definition of a partial system breakdown can be that it is a state, where one or several extra generators or high voltage direct current (HVDC) links tripped due to the extended fault duration.

This additional state may be useful since it indicates a case where generation reserves are needed to keep the frequency inside the limits or the power transmission reserves are needed to keep the power flows under the acceptable values.

7.4 Combination of the Substation Model and Grid Simulations

This section describes how the substation model and the dynamics simulations are combined and analysed, i.e., how the dynamic simulation results are added as new consequences to event trees.

7.4.1 Power System Consequences in the Event Trees

Figure 6.1 presents a very simple event tree with both the *substation* and *power system* consequences added to the end branches of an event tree. Figures 6.7 and 6.8 present the event tree of the PSA substation model after line faults with substation consequences only after a line fault in the middle section of the line. The substation consequences are always the same and not dependent on the power flow. The frequency of different consequences varies to some extent from line to line and is dependent on the line length, substation arrangements and components used at the substations. If two different lines in different locations were as long and had similar components in the substations, the substation consequences would be similar.

The power system consequences vary according to the power flows. If here, too, two different lines in different locations were as long and had similar components in the substations, the power system consequences would not necessarily be similar; faults near big generators tend to have more severe consequences as the fault duration increases. The fault duration increase affects less to stability far away from big generators. This means that the protection system and circuit breaker failures after the faults of different lines lead to different power system consequences. The power system consequences are also a function of the power flow and grid topology. The same substation consequence can in one power flow lead to a critical safe operation state and in some other load flow it can lead to a system breakdown. Therefore, dynamic simulations were made for defining the power system consequences at different grid topology and load flows.

After all the dynamic simulations are made and the consequences are classified, the consequences can be added to the event trees. For example, always when an end branch of a certain event tree leads to a system breakdown, this power system consequence is added to the corresponding end branches of the event trees.

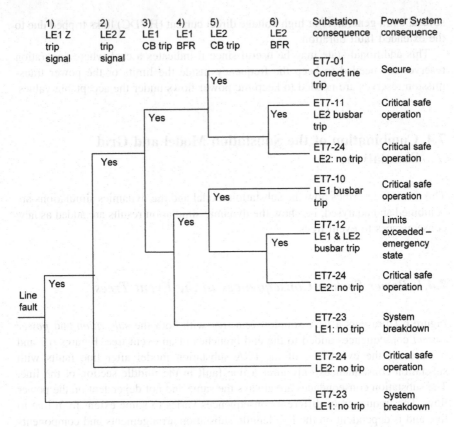

1) LE1 Z trip signal	2) LE2 Z trip signal	3) LE1 CB trip	4) LE1 BFR	5) LE2 CB trip	6) LE2 BFR	Substation consequence	Power System consequence

ET7-01 Correct ine trip — Secure

ET7-11 LE2 busbar trip — Critical safe operation

ET7-24 LE2: no trip — Critical safe operation

ET7-10 LE1 busbar trip — Critical safe operation

ET7-12 LE1 & LE2 busbar trip — Limits exceeded – emergency state

ET7-24 LE2: no trip — Critical safe operation

ET7-23 LE1: no trip — System breakdown

ET7-24 LE2: no trip — Critical safe operation

ET7-23 LE1: no trip — System breakdown

Fig. 7.8 The event tree with power system consequences added to the end branches

Figure 7.8 presents an event tree, similar as in Fig. 6.7, now with power system consequence added. In this case, the power system state is a system breakdown if the trip at LE1 (line end 1) is totally missing due to protection failure or due to the failure of breaker failure protection. Other delayed trips lead to a critical safe operation state.

After all the power system consequences received from dynamic simulations are added to the end branches of the event trees of all lines, the consequence analysis of the system breakdown and partial system breakdown for the whole transmission grid can be done. This is described in Chap. 8.

Reference

1. Nordel (2007) Nordel's Grid Code 2007. http://www.entsoe.eu/index.php?id=62. Accessed 13 May 2010

Chapter 8
System Level Results

8.1 Introduction

In this chapter, the combination of the substation model and power system simulations is described. This chapter discusses the results of the event tree analysis; the chains of events, minimal cut sets, contributing factors, importance measures and sensitivity analyses are presented. Furthermore, based on these results, some recommendations are made. This chapter also includes a description of indices to the system breakdown.

A power system state as a common measure enables importance measures in such a way that different components can be ranked. The frequency of the power system states can be estimated; but more important than this are the minimal cut sets and chains of events that may lead to a system breakdown or other power system states. Different importance measures, Fussell-Vesely, risk achievement worth (also called risk increase factor), risk reduction worth (called risk decrease factor, too), are presented.

Similar devices at different locations of the power system affect the system in different ways, but a systematic analysis can reveal the quantitative differences. The importance measures of basic events (substation components) for the whole grid are presented. The model of this kind provides the structural importance of the components if failure rates are put equal. This is beneficial if the exact failure rates are not known.

It is worth noting that the figures and results presented here strongly correlate to the example presented in this book: the Finish 400 kV intact grid in year 2003, one power flow only; only the line shunt faults that can be tripped by the distance relays and the assumptions and choices presented in Chaps. 6 and 7. The purpose of the example is to illustrate the method and its applicability.

L. Haarla et al., *Transmission Grid Security*, Power Systems,
DOI: 10.1007/978-0-85729-145-5_8, © Springer-Verlag London Limited 2011

8.2 Results of Event Tree Analysis

8.2.1 Grid Level Results

After all the event trees are built, grid simulations made and the consequences of the grid simulations are added to the end branches, the event trees are analysed.

When calculating the contribution of the failures of one line to the power system state, the initiating event frequency of each line fault is needed. All the event tree analyses are here done in such a way that a line fault (initiating event) has a certain frequency. In this example, the initiating events are line faults and it is assumed that the average annual line fault frequency per line length is constant. The initiating event frequency, therefore, depends on the constant line fault frequency and on the line section length and is calculated according to Eq. 6.1. The calculation of the line fault estimate is presented in Sect. 4.3.

In the event tree analysis, the goal is to get the grid level results. The minimal cut sets, importance measures, the frequency of a system breakdown and other power system states can now be calculated for power system consequences instead of local substation consequences. The consequence analysis results are, therefore, at the grid level. The following indexes can be calculated from the event tree results:

- The frequency of a system breakdown after a fault at a line.
- The frequency of a system breakdown assuming the power flow would remain the same. If several power flows are simulated and their relative durations included, an estimate for the total system breakdown can be calculated, too.
- An index to the relative importance of each component in relation to system breakdown and partial system breakdown. This can be done by ranking the grid level importance measures.
- Different probability values of each fault tree can be obtained.
- Local indices to each initiating event can be calculated from each event tree.

To illustrate the PSA method, the frequencies of a system breakdown, and the corresponding minimal cut sets are presented. Also Fussell-Vesely importance, risk achievement worth, risk reduction worth, and the sensitivity of parameters are presented. After the grid level importance measures for all event trees have been calculated, they can be arranged so that the most important substation components contributing the system breakdown after the line short circuits can be recognised.

Sensitivity analyses are used to show how sensitive the results are for the variations of different parameters. Parameters in the event tree model are, for example, the failure rate, test interval and unavailability. The sensitivity of a parameter indicates the rate of change of the consequence if the parameter changes.

8.2.2 System Breakdown Frequency

The frequency of the system breakdown after line shunt faults is a result of the (simultaneous) analysis of all event trees. The frequency of the system breakdown $f(\mathrm{SB})$ is the sum of the individual system breakdown frequencies $f(\mathrm{SB})_k$ of each event tree k and is defined by

$$f(\mathrm{SB}) = \sum_{k=1}^{M} f_k(\mathrm{SB}), \tag{8.1}$$

where $f_k(\mathrm{SB})$ is the system breakdown frequency of an event tree k and M is the total number of the line sections (and event trees).

The estimates for partial and total system breakdown frequencies due to failures at the substation after line shunt faults with the Finnish data were made by Pottonen [1]. The simulations were made for a lightly loaded grid only, which means that the values were, to some extent, too optimistic. The results would be relevant if the power flow was always light.

The estimate was $1.37\mathrm{E}{-}03$ year^{-1} for the system breakdown and $1.12\mathrm{E}{-}01$ year^{-1} for the partial system breakdown. The corresponding time intervals between the successive breakdowns are 730 and 9 years, respectively. Here, a partial system breakdown means that extra components, such as a generator or a high voltage direct current link trips due to extended fault duration. The estimate for annual line shunt fault frequency used in the analysis was $2.9\mathrm{E}{-}03$ faults per kilometre.

There were two different series of events that led to a system breakdown. The most common cause was the failure to trip at the substation, after which the reach of the remote backup protection system was not sufficient to isolate the fault. This kind of series of events caused a system breakdown after faults at 26 of 39 lines. The substation consequences that caused this system breakdown are numbered as 23 or 24 and are presented in Figs. 6.7 and 6.8 and Tables 6.2 and 6.3. The system remained dynamically stable in the simulation time (20 s), but was classified as a system breakdown due to the insufficient reach of the remote backup distance relays. (In reality, if there is zero sequence current, a sensitive earth fault relay can trip the fault after a delay, for example 3 s, but it is possible that the stability would be lost before that. The operation of sensitive earth fault relays was not modelled.)

The other, and significantly less frequent, cause that resulted in a system breakdown was delayed fault clearance near the generators. The extended fault duration was caused by the circuit breakers that failed to trip after a trip signal or by the telecommunication channel failure that caused the delay in the trip signal. This was the case after faults at six lines. The extended fault duration in these cases was either 250 ms (stuck circuit breaker) or 450 ms (telecommunication failure). If a circuit breaker fails, the fault duration is 250 ms and several lines are tripped in one-breaker substations. If the telecommunication fails, the faulted line is tripped after 450 ms, a long enough delay to cause the loss of transient stability of the system.

Seven lines were such that there were no system breakdowns after the fault sequences studied.

The results reveal that after faults in many locations, the system can remain stable, even though the main protection systems or circuit breakers fail. Possible other consequences of faults with a long duration are not calculated since the model is limited to stability. The results also tell that since the backup protection systems do not completely cover the adjacent lines in many locations, there is a threat to security. A significant result is the list of fault locations, where the extended fault duration causes a system breakdown. This means that the inspection and tests of the protection systems and circuit breakers at the substations connected to these lines are more important than similar components at other substations. These results were for a light power flows only. It is evident that with increasing power flows the number of the fault locations leading to system breakdown if not correctly tripped increases.

8.2.3 Minimal Cut Sets

For the whole grid, the system breakdown frequency is 1.37E−03 and the corresponding number of the different minimal cut sets is 13,963. This comes from a model with 39 lines. The results show that 100 minimal cut sets with the highest frequencies represent 81% of the total system breakdown frequency. This means that 13,863 cut sets together respond 18.9% of the system breakdown. Figure 8.1 presents the frequencies and components of the most important cut sets. It appears that minimal cut sets consisting of two failures of a circuit breaker represent about half of all the minimal cut sets. Here, the circuit breakers are not duplicated, i.e. there are no circuit breakers in series since they are expensive components. The circuit breakers that most contribute the system breakdown are air-blast circuit breakers.

The minimal cut sets for '*the total failure to trip*' always have two components. These components are either two circuit breakers at the single circuit breaker substation, two main protection relays protecting the same line, or the telecommunication of the main protection 1 and the relay of the main protection 2.

Figure 8.2 shows the frequencies and components of the 21 most important cut sets, contributing some 47% of the total system breakdown frequency.

Table 8.1 presents the 21 most important minimal cut sets and fault locations. The numbers and contributions of minimal cut sets in Figs. 8.1 and 8.2 and in Table 8.1 illustrate how a PSA analysis of this kind helps to identify the most important fault locations and component failures. Even though there would be tens of thousands combinations of grid fault and component failures, a limited number of combinations may represent a large share of all the minimal cut sets. In this way, the results can reveal the vulnerable parts and components of the system.

At a few fault locations, the power system experienced a system breakdown due to transient stability. The minimal cut sets representing this have one basic event

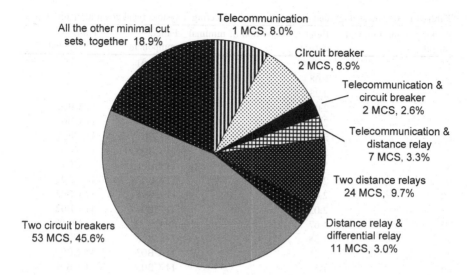

Fig. 8.1 The frequency contributions of components of 100 most important minimal cut sets (MCS) for the consequence *system breakdown*. The 100 minimal cut sets represent 81% of all minimal cut sets, all the remaining cut sets cover 18.9%

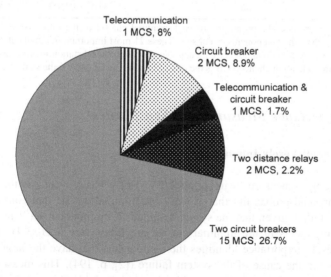

Fig. 8.2 The frequencies and components of 21 most important minimal cut sets (MCS) of a system breakdown. These 21 minimal cut sets represent 47% of the total system breakdown frequency

only; it is either one circuit breaker or one telecommunication channel. These components were the highest in the minimal cut set ranking and are ranked high in all importance measure lists, too.

Table 8.1 The most important minimal cut sets affecting a system breakdown after line faults

Ranking of the minimal cut set	Relative proportion of minimal cut set	Failed component 1	Failed component 2
1	7.96%	TELE 01	
2	6.68	25 CB05	
3	2.23	25 CB05	
4	2.06	11 CB04	11 CB01
5	2.06	25 CB05	11 CB06
6	2.06	11 CB03	11 CB01
7	2.06	11 CB06	11 CB04
8	2.06	11 CB01	11 CB02
9	2.02	11 CB03	11 CB02
10	2.02	11 CB01	11 CB02
11	2.02	11 CB04	11 CB02
12	1.67	20 CB 16	TELE 02
13	1.58	11 CB03	11 CB02
14	1.58	11 CB03	11 CB04
15	1.58	11 CB03	11 CB01
16	1.41	25 CB12	25 CB07
17	1.41	25 CB07	25 CB14
18	1.39	26 CB08	26 CB11
19	1.39	26 CB08	26 CB09
20	1.13	32 Z STA 01	32 Z STA 02
21	1.06	32 Z STA 03	32 Z STA 04

The table also presents the proportion of each minimal cut set of the total system breakdown frequency and lists the components of cut sets. These 21 most important minimal cut sets present 47% of the total system breakdown frequency. *Numbers* are the substation identifications, TELE is the telecommunication channel, CB is the circuit breaker, and Z STA is the static distance relay

8.2.4 Importance Measures and Sensitivities

8.2.4.1 Fussell-Vesely Importance

Fussell-Vesely's measure of importance FV(i) of a basic event i is the approximate conditional probability that at least one minimal cut set that contains component i is failed, given that the system is failed. A minimal cut set is failed when all the components in the minimal cut set are failed (see Sect. 3.4). Thus, the Fussell-Vesely importance identifies the components that have the largest probability of being the cause of the system failure ([2], p. 194). This measure is also called the fractional contribution of a basic event to risk [3]. It is a positive number between 0 and 1. The Fussell-Vesely importance of the component i for a system breakdown is

$$I_{SB}^{FV}(i) = \frac{P(D_i)}{P(SB)} \qquad (8.2)$$

where $P(D_i)$ is the probability that at least one minimal cut set that contains component i is failed, and $P(SB)$ is the probability that the system is failed.

8.2.4.2 Risk Achievement Worth

Risk achievement worth (also called Risk increase factor) is the ratio of the (conditional) system unreliability if component i is not present (or is failed) over the actual system unreliability. It presents a measure of the worth of component i in achieving the present level of system reliability and indicates the importance of maintaining the current level of reliability for the component ([2], p. 191). For coherent systems (see Sect. 4.3.6), the risk achievement worth is always greater than 1. The value of risk achievement worth of a basic event i for a system breakdown SB is

$$I_{SB}^{RAW}(i) = \frac{P(SB|(q(i) = 1))}{P(SB)} \tag{8.3}$$

where $q(i)$ is the unavailability of basic event i. The risk achievement worth can be calculated as a function of Fussell-Vesely importance. In this case, it is calculated as

$$I_{SB}^{RAW}(i) = \frac{1}{1 - I_{SB}^{FV}(i)} \tag{8.4}$$

The failure frequency of the system increases if component i is removed from the system (assuming that the system is coherent). After removal, the new system failure frequency is the product of the original system failure frequency and risk achievement worth of component i.

8.2.4.3 Risk Reduction Worth

Risk reduction worth (also called risk decrease factor) is the ratio of the actual system unreliability over the conditional system unreliability if component i is replaced by a perfect component ([2], p. 191). The risk reduction worth identifies the basic event that would improve the system most if it were perfectly reliable. For coherent systems (see Sect. 4.3.6), the risk decrease factor is always greater than 1. The risk reduction worth of a basic event i for system breakdown SB is

$$I_{SB}^{RRW}(i) = \frac{P(SB)}{P(SB|(q(i) = 0))} \tag{8.5}$$

where $q(i)$ is the constant unavailability of basic event i. If risk reduction worth of component i is 2, then the system unreliability would be 50% of the original unreliability if component i was replaced with a perfectly reliable component.

8.2.4.4 Important Components for a System Breakdown

The results of event tree analysis are ranked according to Fussell-Vesely importance in Table 8.2. Only the components having Fussell-Vesely importance values higher than 0.01 are listed, that is to say, the table has only 32 components listed. Amongst the most important components according to Fussell-Vesely are 18 circuit breakers, 11 distance relays and 3 telecommunication channels. Most, but not all, circuit breakers are air-blast circuit breakers. The most important components in Fussell-Vesely ranking are four air-blast circuit breakers. Their FV measure varies between 0.11 and 0.13. The ranking lists of the most important components according to Fussell-Vesely and risk reduction worth are similar to each other as Table 8.2 shows. This is natural since risk reduction worth importance identifies the basic events that would improve the system most if it were

Table 8.2 The most important components ranked according to their Fussell-Vesely and risk reduction worth importance

Ranking according to Fussell-Vesely and risk reduction worth	Component identification type	Fussell-Vesely importance (FV)	Risk reduction worth (RRW)	Risk achievement worth (RAW)
1	11 CB01	0.132	1.15	16.4
2	11 CB02	0.131	1.15	16.2
3	11 CB03	0.119	1.14	14.8
4	11 CB04	0.110	1.12	13.8
5	25 CB05	0.093	1.10	66.0
6	TELE 01	0.080	1.09	16.8
7	11 CB06	0.064	1.07	8.48
8	25 CB07	0.047	1.05	6.51
9	26 CB08	0.046	1.05	6.42
10	26 CB09	0.045	1.05	6.30
11	11 CB10	0.043	1.04	5.97
12	26 CB11	0.035	1.04	5.14
13	25 CB12	0.032	1.03	4.73
14	26 CB13	0.030	1.03	4.52
15	25 CB14	0.030	1.03	4.45
16	26 CB15	0.028	1.03	4.21
17	TELE 02	0.026	1.03	1.05
18	TELE 03 B	0.023	1.02	2.87

The list also shows the risk reduction worth and sensitivities of the components. *Numbers* are the substation identifications, CB is the circuit breaker, TELE is the telecommunication channel

Table 8.3 The components ranked the highest according to their risk achievement worth importance

Identification	Risk achievement worth
13 VT MCB of electromechanical relays Z01 and Z02	341
25 VT MCB of electromechanical relays Z07 and Z08	246
26 VT MCB of electromechanical relays Z09 and Z10	242
13 VT MCB of electromechanical relays Z03 and Z04	242
13 VT MCB of electromechanical relays Z05 and Z06	187
26 VT MCB of electromechanical relays Z11 and Z12	144

This list presents only single components with RAW greater than 100, but not any common cause failures. The *first numbers* are the substation identification codes, VT MCB is the miniature circuit breaker of a voltage transformer, common for two electromechanical distance relays

perfectly reliable and Fussell-Vesely identifies components that have the largest probability of being the cause of the system failure.

The most significant components according to risk achievement worth were different from Fussell-Vesely and risk reduction worth. The basic event representing common cause failures *substation* and *bay* had the highest risk achievements worth importance values. These basic events were for common cause failures and a basic event of a certain bay or a certain substation is in all the fault trees of that bay and substation, respectively. In addition to this, all the miniature circuit breakers of electromechanical distance relays were ranked high in the list of risk achievements worth importance. The two electromechanical distance relays protecting the same line have a common miniature circuit breaker in the voltage transformer circuit. If the miniature circuit breaker trips, both relays are incapable of tripping the line, and this property caused these devices a high ranking when looking at the risk increase if the component would be disconnected or always failed (Table 8.3).

8.2.4.5 Sensitivity of Parameters

Parameters in the event tree model are, for example, the failure rate, the test interval and constant unavailability. The sensitivity of a parameter indicates the rate of change of the consequence if the parameter changes. The sensitivity S_h of any parameter h for the system breakdown is often calculated

$$S_h(\text{SB}) = \frac{P(\text{SB}|10h)}{P(\text{SB}|0.1h)}. \tag{8.6}$$

The sensitivity of the results of parameters tells how much the results would change if the value of the parameter changed. The model parameters were failure rates, test intervals, repair times and, for some components, unavailability values.

Table 8.4 The parameters with the highest sensitivity values

Component	Parameter	Original value	Sensitivity
Circuit breaker, all types	Test interval	1 year	148
Air-blast circuit breaker	Failure rate	1.72E−02 faults/year	77.2
Microprocessor distance relay	Unavailability	3.3E−03	7.53
SF6 circuit breaker	Failure rate	2.8E−03 faults/year	6.53
Static distance relay	Unavailability	7.1E−03	5.75
Minimum oil circuit breaker	Failure rate	4.9E−03 faults/year	5.62
Electromechanical distance relay	Unavailability	3.3E−03	3.07

The ranking list of parameter sensitivities in Table 8.4 shows that the circuit breaker testing interval and the failure rate of air-blast circuit breakers are the parameters that have the highest sensitivity values. In the model, the circuit breaker test interval was 1 year because the relays are tested annually in such a way that the circuit breaker trip after it has received a trip signal is part of the test. The results indicate that it is important to check the operation of the circuit breakers at least once a year even though the relays would be tested less frequently. This list ranked also quite highly some unavailability values of the distance relays.

8.3 Comments and Recommendations

A PSA model should be regularly updated according to changes in the system and new failure data. After updates, the results may be different. Each result is a snapshot of a situation. The numbers in the results and different frequencies are only a part of the results. The main results are the vulnerabilities that the results show. The PSA method helps to reveal the complexities of the system vulnerabilities and the connections between the faults, component failures and power system dynamic behaviour.

Some equations, for example Eq. 5.6, are not correct. Equation 5.6 is based on the assumption on only periodical inspections that reveal failures. If some failures are detected during disturbances or when the component is needed, the unavailability becomes smaller that the equation provides. There is another error in the assumptions behind the equations of unavailability values. The repair is assumed to change the state of a failed component to as good as new, which is not the case, if only a part is repaired.

Also the assumption of a constant failure rate is not correct for all devices. Often the infant mortality is high, and the failure rate starts to increase when the component becomes old. The constant failure rate is valid only during the period after the infant mortality but before the failure rate increases significantly.

8.3.1 Importance and Sensitivity

The Fussell-Vesely importance is directly proportional to the unavailability value of the component. Thus, Fussell-Vesely importance measures can be used alone for identifying the potential components for safety improvement. The risk reduction worth is identical to Fussell-Vesely in this respect. The risk achievement worth measure, on the other hand, is a weak function of the unavailability of the component, and thus, it sees the system from a different point of view. It does not represent the component itself, but the defence of the rest of the installation against a failure of a component [4].

The Fussell-Vesely importance measures at the system level, presented in Table 8.2, show that the failure of a circuit breaker is in many fault locations the cause of a system breakdown. Many, but not all, of these circuit breakers were old air-blast circuit breakers. Some circuit breakers near big generators had a high ranking in Fussell-Vesely's lists, because the extended fault duration due to a stuck circuit breaker followed by the weakening of the grid due to a busbar trip (by a breaker failure relay) caused the generator to lose their synchronism and finally led to a system breakdown due to instability. Some circuit breaker failures near the power plants caused generator trips, even though the system remained stable. The parameter, the change of which most changed the system breakdown frequency, was in many fault locations the circuit breaker test interval.

As the circuit breakers were often the main reason for grid problems, the circuit breaker test interval should not be lengthened. The test interval of the circuit breakers in the model was 1 year since that was the period between regular relay tests when the breakers were checked, too. The high sensitivity of the system for the circuit breaker test means that the circuit breakers should still be checked once a year even though the microprocessor relay test interval would be increased due to the self-supervision they have.

The above example shows that the air-blast circuit breakers should always be changed to SF6 circuit breakers when the substation is renovated. At the substations near the generators, the circuit breakers could be changed even without the substation renovation.

8.3.2 Protection Systems

The single components, the failing of which most increases the risk of a system breakdown, are the miniature circuit breakers of the electromechanical distance relays. After the old relays are exchanged for new ones, both relays will have their own miniature circuit breakers. Thus, this safety problem will disappear in the future.

As some telecommunication channels had a high ranking in the Fussell-Vesely ranking, a good way to increase the security is to duplicate all the telecommunication channels. The doubling changes the system to a more balanced one, since after the duplication no single component can represent almost 8% of the total system breakdown frequency. After the analysis, all the line protection telecommunications were duplicated.

The reach of the third zone of the distance relays is sufficient only on short lines. On many locations, it is not possible to lengthen the reach to all fault locations of the adjacent lines since the third zone should never reach behind the adjacent lines. It is worth noting that a lengthening of the reach can increase the risk of unwanted trips.

A good method of improving the reliability of duplicated protection systems would be to have such specifications that the main protection relays were obtained from different manufacturers. Then the possible common cause failure possibility due to the same software code in the duplicated relays would disappear.

8.3.3 Other Results

Slightly different conclusions can be drawn from the results of the substation reliability model alone, without a consideration of the power system impacts. The relays that cause the failures at the substation more often are the static distance relays compared with electromechanical or microprocessor relays. The same comparison between the circuit breakers gives the result that air-blast circuit breakers fail more often than minimum oil and SF6 circuit breakers. Here it is worth remembering that the blocking of SF6 circuit breakers due to low gas pressure is not used.

In a similar way, different protection systems for busbar and the line faults can be compared in order to see the differences between the relays or telecommunication channels. It is possible to compare the effect of duplication of components on the reliability.

During the modelling process, a lot of useful information can be achieved. The qualitative results, too, can give information on the system. The failure mode and effect analysis (FMEA) process can give good hints of how to keep the failure statistics of components. The more detailed the failure statistics are, the better the models can be used for PSA: Often primary components, such as circuit breakers, have more detailed statistics than secondary components, such as the relays. It is good to include component years in statistics. If only failures are listed, it is not possible to calculate failure rate estimates. It is also possible that during the modelling, some flaws in figures, specifications or installations can be found. As a matter of fact, an incorrectly made duplication of a telecommunication channel was found and corrected when the PSA model for the Finnish grid was applied.

The results received with the analysis of the transmission grid give an overview of the different series of events that lead to system problems. They also present the relative importance of different components as a reason for a system breakdown or a partial system breakdown.

References

1. Pottonen L (2005) A method for the probabilistic security analysis of transmission grids. A doctoral dissertation, Helsinki University of Technology, 951-22-7591-0, 951-22-7592-9. http://lib.tkk.fi/Diss/2005/isbn9512275929/. Accessed 29 June 2010
2. Rausand M, Høyland A, (2004) System Reliability Theory, Models, Statistical Methods, and Applications. Second Edition, John Wiley & Sons, Inc. Hoboken, New Jersy. ISBN 0-471-47133-X
3. Mankamo T, Pörn K, Holmberg J (1991) Uses of risk importance measures. VTT Research Notes 1245. ISBN 951-38-3877-3. ISSN 0358-5085. Espoo Finland
4. Van der Borst M, Schoonakker H (2001) An overview of PSA importance measures. Reliab Eng Syst Saf 72(3):241–245

Chapter 9
Perspectives for Future Power System

9.1 General

The chapter describes the framework to ensure that secure system operation is also met with the electricity market and in the future power system. Legislation and agreements set requirements for transmission system planning and operational planning. The market design takes care of the implementation. In this section, the European electricity market design is referred. The European market design illustrates ways for implementation and possibilities to face challenges introduced by electricity markets, the sustainable production of electricity and new emerging technologies.

Section 9.2 briefly describes electricity markets, sustainability and emerging new technologies in relation to system reliability. Section 9.3 addresses the cooperation and coordination between transmission system operators when network planning spans across several control areas within integrated electricity markets. Section 9.4 reflects the requirements for operational cooperation and common rules for operation and grid connection within interconnected systems and markets for system security. Section 9.5 discusses the different congestion management methods to ensure that induced power flows are within transfer limits. Section 9.6 concludes the challenges, which the PSA approach faces with more tightly interconnected power system and possible future changes in the dynamics of the power system.

9.2 Introduction

9.2.1 Electricity Market

The sale of electric energy faces competition in the electricity market, whereas transmission or distribution is considered as a natural monopoly and is regulated accordingly. Due to the scale of economies, these regulated natural monopolies operate more efficiently than with a competitive transmission market. All functions belonging to production and supply of electric energy has to be separated either legally or through accounts from transmission and distribution to ensure that no cross-subvention between competitive and monopoly business functions occurs.

Electric energy can be sold in several ways:

- bilaterally through an OTC (over the counter) market,
- in organised markets within a power exchange or a pool.

In the bilateral trade, the market actors make an agreement to sell and buy a fixed physical block of electric energy for a fixed period of time. Because trade is physical in nature, the place of delivery will also be defined in the agreement. Bilateral trade may occur either within national borders or across the borders. Trade occurring within national borders may be applied without any restrictions because there is enough transmission and distribution capacity to meet the needs of the traded electric energy. However, if not enough transmission capacity exists between output and intake places, then market actors should also take care of the transmission capacity reservation for the bilateral trade between the output and intake places. The scarcity of transmission capacity often appears on the cross-border interconnections. This implies that market actors need to reserve transmission capacity for cross-border bilateral trade, and trade may be denied if no adequate transmission capacity exists.

Market actors often favour bilateral markets because there is possibility for increased income through commissions and price spreads. Bilateral trades are also applied commonly in longer time frames, e.g. in annual or multiannual time frame. Bilateral markets are quite flexible, and market participants may agree on any contract they want. The flexibility has the price as negotiations, and making agreements may take time, and counter party risk may be substantial. For these reasons there has emerged need for more organised trading aiming at increased volumes and liquidity.

The organised trading is executed usually through power exchanges or pools. The power exchange is a market place, where market actors can sell and buy either financial or physical electric energy. These market places are usually established voluntarily to meet the needs of the market actors. The market actor has to meet the requirements and rules set by the power exchange. An exchange acts as a counter party for all trades, thus decreasing the counter party risk of bilateral trade. The product portfolio of the power exchange is usually aligned to meet the needs of the market actors implying both financial and physical products of different duration and volume. Financial products have usually the duration of some days until a couple of years. Physical products are sold quite near the real time, e.g. from a day-head time frame until an intra-day time frame.

The organised markets can be established also as pools. Usually, these pools are established as mandatory pools implying that all trade within a specific geographical area will occur in the pool.

The power exchange introduces some benefits compared to the bilateral trade. The power exchange may increase competition, be transparent in price formation and reduce trading costs. Furthermore, the power exchanges are considered operating much faster than bilateral markets as power exchanges can execute trade within a few minutes, while it may take hours to negotiate and agree on the bilateral contracts. For power trading, this rapidity is crucial, and thus power exchanges have been established to manage trade near the real time, e.g. within day-ahead and intra-day time frame.

9.2.2 Sustainability

Prevention of the climate change requests sustainable energy production. To globally meet the target to decrease the CO_2 output from the production of the electric energy, for example, the EU member states have agreed to increase the amount of electric energy produced using the renewable energy sources to 20% by the year 2020. The national targets within the EU member states have been set to meet the general 20% EU target taking into account the member states' possibilities to harness new renewable resources.

Renewable energy resources applicable to electric energy production include hydro, wind, solar, biomass, geothermal and wave energy. Prominent to most of these renewable energy sources is their intermittency or variability, i.e. electric energy is produced only when wind is blowing or sun is shining. The effects of this variability have to be taken into account in the evaluation of the power system security when the amount of renewable production increases. The application of renewables asks for large interconnected power systems to adapt to the variable power injections from the renewables. A large power system also has large geographical spread, implying that wind conditions and wind energy production are most probably different across the system. The wind energy produced in places having stronger wind at certain moment can be applied for consumption in places where the wind energy production does not exist at the same moment due to the non-existent or weak winds. This wind variation across the large power system presumes that the connections between and within the power systems are strong, and emerging power flows do not exceed the transmission capacity of the power system and lead to congestions. These variations introduce more uncertainty for system reliability assessment, and the applied methods should cope with these variable power flows.

This renewable production can be realised with distributed generation (or embedded generation) or large remote renewable production parks, e.g. off-shore wind parks or on-shore photovoltaic parks. Distributed generation is small size generation located near (or within) a consumption site. This implies that distribution networks have to securely adapt this embedded generation. Large off-shore wind parks instead require a strong grid to transport produced energy securely to the consumption sites located far away from these production sites.

Fluctuations in renewable production request more power reserves to maintain system frequency within allowed limits in order to avoid the disconnection of conventional large generators due to low frequency. The effects of these frequency fluctuations should be considered in the security analysis.

The electricity market design has to be amended to better adapt to the characteristics of renewable production, e.g. by introducing a market nearer to real-time operation to thus enable owners of renewable generation the possibility to change their positions accordingly. Otherwise, transmission system operators have to adapt larger balance between demand and supply during the operational hour by the balancing market. This may in some situations, i.e. if there is lack of bids in the balancing market, violate the security of the power system.

9.2.3 Emerging Technologies

New generation technologies will be introduced for renewable energy sources. Static converters will be applied increasingly for converting primary energy to electric energy instead of prevailing synchronous generators. This will change the characteristics and dynamics of the future power system and should be taken into account in the security planning. Introduction of large amounts of distributed generation calls for increased monitoring and control to maintain the secure system operation. This requests reliable telecommunication and control systems to ensure security. The effects of these systems should be modelled in the security analysis.

More direct current (DC) links will be introduced to transmit power from off-shore wind parks to the on-shore substations. Prevailing alternating current (AC) grids has to accommodate more these radial direct current links or even meshed off-shore direct current grids. This development introduces another complexity for security analysis, where the interaction between controls of direct current links and alternating current power systems has to be considered.

The efficient utilisation of the grids implies that the number of monitoring and controlling equipment increases. Wide area monitoring systems (WAMS) will be applied. The remote metering of consumption sites with automatic disconnection possibilities may contribute to the automatic demand response when system frequency is in danger.

The future power system is more dependent on the information and communication technologies (ICT). The reliability of these technologies and interactions with power systems should be modelled when the reliability of the future power system will be considered.

9.3 Planning of Interconnected Power Systems

The planning of interconnected power systems calls for harmonised processes, congruent assumptions and security criteria for the network planning. The national

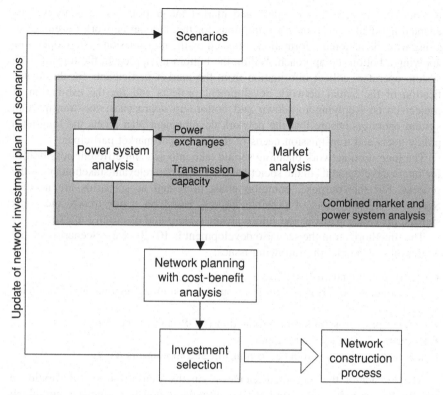

Fig. 9.1 General network planning process

planning process should be adapted to cover the interconnected power systems and markets and to be enlarged to a common network development plan across relevant transmission system operators. This ensures that the grid is built according to the needs of the integrated markets.

To harmonise the planning process, the transmission system operators should agree on the time period to prepare the common network development plan: annual, biannual or longer time period. Usually these common network development plans are prepared within a longer time period (e.g. every second year), and national plans may be updated within a shorter time period (e.g. annually). Furthermore, the transmission system operators should agree on the time horizon for planning. Generally, the planning horizon is from 5 to 10 years. Longer time horizons may be applied for general overview purposes. The general network planning process is presented in Fig. 9.1. The planning process should be consistent for both national and interconnected power systems.

The common network planning process requires both top-down and bottom-up approach to ensure congruent assumptions among relevant transmission system operators. The top-down approach ensures that the common goals for the future are fed in the network planning process. These goals will address the issues related to policies, the security of supply and market integration. These goals will be adapted nationally to produce assumptions for the national network development along with the national information collected by the transmission system operators applying a bottom-up approach. Within the bottom-up approach, the transmission system operators collect information from the market participants for the identification of the actual network development projects and for the estimation of project costs. Applying top-down and bottom-up approaches, the transmission system operators ensure that the network development plan meets the long-term policy goals and transmission needs of the individual market participant.

The long-term network planning should take into account different possibilities for future developments in the society and involve also relevant stakeholders in the process. For this purpose, common goals presented as scenarios are needed. Scenarios are snapshots of possible futures, based on major trends and weak signals, providing paths to the future power system and electricity market.

The time horizon in the scenario development is 10–20 years. Scenarios should address, for example, the following issues:

- energy policy and sustainability,
- the availability of primary energy resources at national, regional and worldwide level,
- demography, gross domestic production (GDP) and financial resources,
- the price of primary energy, and
- new technologies to produce, to transmit and to consume energy.

The impact of the identified scenarios on the transmission grid should be studied by a model consisting of the integrated power system. A combined model of electricity market and the power system should be used to determine the transmission needs required by each scenario. The market model should include all relevant parameters having an influence on the integrated electricity market. Within a scenario, the market parameters define the balance between supply and demand across the integrated market. The duration curves for power exchanges between the identified demand and supply areas are defined together with the costs of congestions if limitations in power exchanges occur. The market model contributes to the evaluation of potential congestions in the transmission capacity.

The transmission capacity of the existing grid and alternative reinforcements in the grid for the identified scenarios under different operating conditions is analysed with a power system model. Power exchanges from the market model are applied to evaluate the needs for transmission capacity in the power system model. The calculations should be executed with load flow and dynamic simulation software and a relevant power system model applying common security criteria. The result of this analysis will be alternative network reinforcements meeting the transmission capacity requested by each identified scenario.

The whole integrated power system should be modelled, not only a national or a part of neighbouring power systems, to ensure that efficient investments are made for integrated electricity markets. Compatibility of analysis tools is essential for a proper conversion of exchanged data between the transmission system operators and between different analysis tools.

The cost-benefit analysis will be applied to find out the most efficient network investment projects among the identified alternative network reinforcements. In this analysis, the costs of all alternative reinforcements are compared against the socio-economic benefits these reinforcements introduce. The outcome of the cost-benefit analysis is a list of preferred investment projects indicating also the building time schedules from a few years up to 10 years. The list of identified investment projects forms the network development plan of the transmission system operator. Investment decisions and network construction projects are established based on the network development plan.

The network development plan needs regular updates. These updates take into account the decided investments of transmission system operators and market actors. Furthermore, changes in the plans of the transmission system operators and the market participants are taken into account when updating the network development plan. Scenarios reflect long-term trends. Thus, it is not necessary to develop new scenarios every time the network development plan is updated.

The coordination and cooperation in the network planning between the transmission system operators of the European Union (EU) member states has been ensured by the legislation [1]. The legislation requires the transmission system operators within their cooperation organisation ENTSO-E to publish every second year the community-wide 10-year network development plan. This plan shall be consistent and coherent with the regional and national network development plans by the transmission system operators.

9.4   Framework for Coordinated Operation Between Transmission System Operators

9.4.1   Introduction

Until recently, the transmission system operators have operated quite independently their national transmission networks due to the weak interconnections across the national borders. It has been possible to develop the security rules for operating these transmission systems taking into account only the national requirements. Increased trade across the borders with more interconnections implies that the transmission networks become more tightly connected also physically in the future. It is no more possible to operate them securely without taking into account the effects of neighbouring transmission networks.

Within a synchronous area, transmission system operators have jointly set rules to operate the interconnected power system. These operation rules aim to support the secure operation of the interconnected networks within a synchronous area and to ensure the interoperability between transmission system operators within a synchronous area. The rules may include requirements for the parties that operate generation and consumption, setting up reserves and defining security criteria including the criteria to define the transmission capacity of the interconnected power system. Also requirements for reporting, performance monitoring and information exchange may be included. These rules have been developed to ensure the secure operation of the synchronous power system. However, these rules are not commonly set binding through the agreements or the legal framework. Any obligations to comply with the rules have not existed; it has been voluntary for the transmission system operators to comply with the rules.

The electricity market extending to several synchronous areas implies that dependence in the operation between these synchronous areas increases. Thus, operation rules for ensuring secure operation should be made compatible across the affected synchronous areas. Operating such large areas with several transmission system operators with voluntary rules is challenging. Solution for this has been to make the operation rules binding either through agreements or through the legal framework. Implementation of these rules is to be monitored to ensure the compliance.

Within the integrated electricity markets, there is still a need for rules that cover only national transmission and distribution networks. These rules have mainly national impact, and thus, they are not relevant in the wider context of integration. These rules may include procedures for network access of customers, billing and balance settlement. However, it is important to ensure that these national rules are compatible with the operational rules covering the whole electricity market area.

Figure 9.2 illustrates the interactions between synchronous power systems, transmission and distribution networks and customers for the coordinated operation between transmission system operators in the integrated electricity markets.

9.4.2 Operation Rules

Security of the power system is vital for the operation of the interconnected power systems. In an interconnected power system, interdependencies between the grids of transmission system operators exist, and operators are not allowed to interfere with the electricity market unless the security of the system is not ensured. Operation rules have been developed for this purpose. The development process should cover the evaluation of security within the integrated power system before any common rule is implemented.

The operation rules address topics like frequency control, common criteria to ensure operational security, coordinated operational planning for outages and information exchange between transmission system operators. Furthermore, training skills for grid operation and coordinated actions during emergencies should be agreed.

During the operational hour, the balance between load and generation should be maintained to keep the system frequency at its nominal value. In the electricity

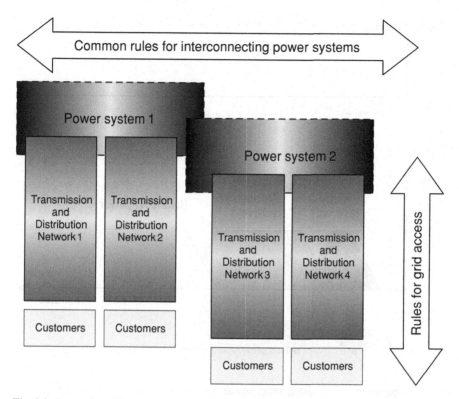

Fig. 9.2 Interactions of power systems, transmission and distribution networks and customers for coordinated operation between transmission system operators

market, the market participants should plan in advance their demand and supply to be balanced during the real-time operation. This implies that generally balance between load and generation will be met in the real-real time operation. However, due to disturbances and forecast errors of load and generation, there is a need for further adjustments to the balance during the operational hour. The transmission system operators have responsibly for these adjustments and to maintain the system frequency within the allowed limits. The transmission system operators within a synchronous area have agreed on the common procedures to efficiently manage the imbalances during the operational hour. For this purpose, the transmission system operators have agreed to have at their disposal a predefined amount of automatically and manually activated power reserves.

Automatically activated reserves are usually generation, where output can be within some seconds or minutes decreased or increased depending on the need. These reserves have been called differently in different synchronous systems, e.g. primary control and secondary control reserves in continental Europe [2] and the frequency controlled normal operation reserve and frequency controlled disturbance reserve in Nordic countries [3]. These reserves are purchased by the

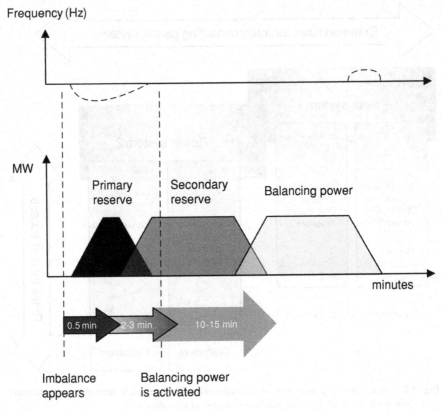

Fig. 9.3 Activation of different reserves to maintain system frequency when imbalance between load and generation occurs

transmission system operators through market-based procedures, e.g. tendering or auctions.

Manually activated reserves will be activated within 10–15 min by the manual instructions from the control centre of the transmission system operator. These manually activated reserves are used to relieve automatically activated reserves for future usage. These manual reserves have been called differently in different synchronous systems, e.g. tertiary control reserves in continental Europe (UCTE 2009) and fast forecast reserve and the fast and slow disturbance reserve in Nordic countries [3]. The manual reserves may be purchased through the balancing power market or through tendering procedure. Figure 9.3 shows the management of the system frequency within the operational hour by different reserves when the imbalance between load and generation occurs. The amounts of these power reserves will be defined to meet the predefined security criteria.

The scheduled power flows across the interconnections should be exchanged between the transmission system operators for the daily operation of the inter-connected power system. These scheduled flows are to be adjusted during the operational hour by the automatic load–frequency control within each control area. However, unexpected deviations may occur in power flows, and the transmission system operators should adapt the operation accordingly to maintain the security.

The operation of the interconnected power systems is based on the principle that each transmission system operator is responsible for the safe operation of its own area and shall coordinate with the neighbouring transmission system operators at a synchronous area level. Common operation rules usually define the scope of cooperation during the real-time operation in order to ease interoperability between the transmission system operators. The interconnections between the neighbouring transmission system operators are the most relevant to the secure interconnected operation, and thus, the most relevant operation rules are related to these interconnections.

A common $N - 1$ criterion has been applied by each transmission system operator to ensure that disturbances do not cause cascading impacts on the neighbouring power systems. Application of the $N - 1$ criterion prevents the propagation of any incident to a larger area. The criterion is valid also for the liberalised electricity market. However, the coordination between the transmission system operators across the synchronous area is a necessity also when developing a common $N - 1$ criterion. A common understanding of the acceptable consequences and of the applied $N - 1$ criterion has been in focus in Europe recently when a fault in the normal operational state caused cascading impacts in the interconnected power system of continental Europe [4, 5].

When applying the $N - 1$ criterion the transmission system operators have to know the consequences that any contingency within their own or neighbouring system may cause to their power system. Coordination between the transmission system operators is an obligation in the liberalised electricity markets because the domestic actions by a transmission system operator may have effect on the neighbouring power systems. Each transmission system operator should inform its neighbouring transmission system operators the application of the $N - 1$ criterion and the security risk level applied. The transmission system operators should also introduce appropriate remedial actions to recover to the normal state after a fault and to avoid cascading events across interconnections after faults. These coordination aspects have to be considered during the coordinated operational planning and in the real-time operation to meet the acceptable security level. Furthermore, these aspects have to be taken into account in the coordinated outage planning and in the transfer capacity assessment.

9.4.3 Grid Access Rules

Each transmission system operator has rules for the access to the network. These rules define the general and specific requirements for the connection to the network. The rules include both physical and commercial aspects of the connection. The rules should ensure non-discriminatory access to the grid users across the integrated electricity market and the security of the interconnected system. The prerequisite in defining the rules for the grid access is the compatibility with the operational rules to ensure the secure operation of the synchronous area and building a level playing field to the market participants within the integrated electricity market.

Within the integrated electricity markets, there is also a need for the grid access rules that cover only the national transmission and distribution networks. These rules have mainly a national impact, and thus, they are not relevant in the wider context. However, it is important to ensure that these national rules are compatible with the operational and access rules covering the whole electricity market area.

The grid access rules create favourable conditions for the cross-border exchanges induced by the network users and by the transmission system operators themselves. The implemented rules ensure that the transmission system operator knows the behaviour of the equipment connected to the grid during the system disturbances. The transmission system operator is thus able to execute the risk assessment of the power system with the relevant power flows.

In this respect, the requirements set for the generators to tolerate disturbances are the most critical. Large generators, especially, are relevant to the behaviour of the whole interconnected system. They should be designed to remain connected to the network during and after nearby faults. This means that their impedances should be low, and they should be equipped with a voltage control to support with the reactive reserves in order to maintain the stability during disturbances. The tripping of generators due to the network faults should be avoided because this increases the risk for the cascading trips of several generators. Generally, the possible tripping of the generators during the system operation is decreased by setting requirements for the fault clearance time exceeding those of the primary protection and requesting that the generators remain connected to the network, i.e. without losing stability, with the longer fault clearance time.

If the transmission system operators set requirements differently for the large generators within a synchronous area, this introduces a risk that the disturbance spreads across the borders. The generators with lower level requirements may not tolerate all the disturbances occurring in the interconnected system, and this may have an effect on the stability of the interconnected system.

9.5 Congestion Management

9.5.1 General Provisions

The commercial transactions by market participants induce power flows in the transmission grid. When power flows caused by the commercial transactions are below allowed power flows, there should not be any restrictions for accessing the grid. In this case, there is no congestion in the grid. In some situations, these power flows may exceed the allowed power flows. These power flows shall be curtailed to ensure secure operation of the power system affecting commercial transactions.

If the commercial transactions induce congestions, transmission system operators shall implement methods for congestion management according to the European legislation. These methods should ensure that the power flows associated with commercial transactions comply with network security standards. Furthermore, transmission system operators should handle congestions economically efficiently. The congestion management methods should give efficient economic signals to market participants and the transmission system operators. The method should promote competition and should be suitable for application in larger geographical area. The method should not be based on a single commercial transaction.

Congestions have effect on the electricity markets while ensuring secure system operation. It is important how the network areas for congestion management are to be defined so that impacts on the markets are minimised and cost-effectiveness is met. Here the congestions within a national system and across the interconnections have to be treated with equal terms.

9.5.2 Congestion Management Methods

The existing European regulations state that methods for congestion management shall be market-based. This implies that transmission capacity shall be allocated only by means of auctions. There are two types of auctions: explicit auctions, where only transmission capacity is sold or implicit auctions, where both transmission capacity and electric energy are sold simultaneously. It is possible to have either one of these methods on the transmission line or both methods on the same line. Here, the bids from market participants, either implicit or explicit, having the highest value shall be executed.

The non-discriminatory treatment of different market participants implies also that market participants having bilateral agreements and those bidding to power exchange shall have equal treatment when congestions occur.

Methods for congestion management may have long- or medium-term transmission capacity allocations, e.g. transmission capacity may be allocated for multi-annual, annual, monthly and weekly time frames. Furthermore, short-term transmission capacity allocations exist, e.g. day-ahead and intra-day time frames. If the transmission line has allocations in different time frames, the principle for allocation in different time frames may be either

- a predefined fraction of the available transmission capacity is allocated to every timeframe with a possibility to allocate remaining capacity not allocated in previous time frames, i.e. one-third of the capacity is allocated to annual actions, one-third of the capacity for monthly auctions and one-third for day-ahead auctions, or
- allocation of all remaining transmission capacity not allocated in previous time frames, i.e. market participants reserve 30% in annual auctions implying 70% for monthly auctions.

Both principles may be adjusted to take into account any transmission capacity released by market participants holding the transmission capacity from previous allocations. It is transmission system operators' duty to define an appropriate structure for the allocation of capacity between different timeframes to meet the requests from the market participants.

The allocated transmission capacity may be either physical or financial transmission right for the capacity holders. The physical transmission right (PTR) gives the holder of the right physical access to the transmission capacity. The financial transmission right (FTR) gives the holder of the right hedge against the price of the transmission capacity. Furthermore, transmission system operators may offer either firm transmission capacity or capacity with reduced firmness. The firmness of the transmission capacity can be either physical or financial. Generally, transmission system operators offer physical firmness in short-term allocations, e.g. day-ahead and intra-day time frames. Financial firmness may be preferable solution in longer time frames.

Market participants may not need the transmission rights allocated to them. To ensure that market participants use the transmission rights allocated to them, they have to nominate the transmission rights at the latest a few days (or 1 day) before the operating day. If they do not use the rights allocated to them, they will either loose the rights, or they have to sell the transmission rights to other market participants in a secondary market. This approach pre-empts the blocking of scarce transmission capacity from those needing the transmission capacity.

The energy transactions are mainly executed in the long- and medium-term time frame and in the day-ahead time frame. The efficient methods for congestion management in these time frames and coordination between the methods in these time frames are important. Here, the common gate closure between different market time frames is also crucial for the efficient functioning of the integrated market.

Market participants need intra-day markets to adjust their forecasts, e.g. for increased intermittent generation or to overcome the outages in generation. The importance of the intra-day time frame is increasing due to the variable generation. The congestion management method for the intra-day time frame should meet the requirements for the fast allocation procedure because allocation happens quite near real time, e.g. 1 h before operational hour.

The balancing of supply and demand in a power system is executed in real time during the operational hour. The transmission system operators are responsible for the balancing, and they shall buy the balancing energy through market-based methods according to the European legislation.

Calculation of available transmission capacity should be consistent across the integrated market. In long- and medium-term, the definition of available transmission capacity may include uncertainties and risks, and thus it is common to take these uncertainties and risk into account when these capacities are defined. In the day-ahead time frame, situation for production, consumption and grid is more predictable, and this implies that amount of available transmission capacity may increase. Furthermore, after the day-ahead market is closed, the uncertainty of grid, production and load situation decreases. This implies that more transmission

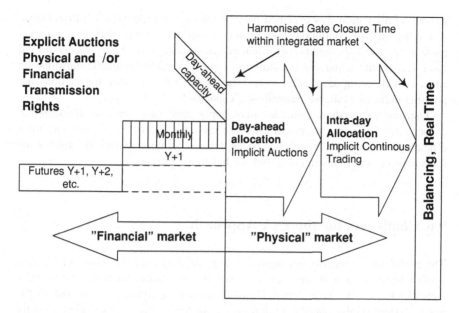

Fig. 9.4 A solution for coordinated allocation of transmission capacity in different time frames

capacity may be available for the intra-day market. Finally, during operational hour, the security check will reveal the actual available transmission capacity at real time.

Figure 9.4 presents one solution to manage congestions within different timeframes [6].

The allocation procedure is divided into two parts, i.e. physical market and financial market. This division occurs in the day-ahead allocation time frame.

In the financial market time frame, the transmission capacity is allocated either by explicit auctions or selling physical and/or financial transmission rights (PTRs, FTRs). These transmission capacity allocations can be multi-annual, annual, monthly or even weekly time frames. These rights should relate to the available transmission capacity. If financial transmission rights (FTRs) are applied, the rights will be cleared in the day-ahead time frame. Application of financial transmission rights enables that all available physical transmission capacity can be allocated in day-ahead and intra-day time frame. If physical transmission rights (PTRs) are applied, the rights have to be taken into account when the physical day-ahead allocation is made. In this case, it might be difficult to define which share of transmission capacity should be allocated in longer term and which share of capacity should be allocated in day-ahead time frame to meet the needs of market actors efficiently.

In a day-ahead time frame, all available physical transmission capacity will be allocated through the implicit auction applying a single price coupling across the integrated market. The remaining physical transmission capacity after the day-ahead allocation was executed may be allocated in the intra-day time frame, where the implicit continuous trading may be applied. In this method, the market participants can bid continuously when no congestions exist. Every successful bid has effect on the remaining transmission capacity, and the available transmission capacity will be updated after every successful bid. Intra-day transactions are possible until all available transmission capacity has been used.

Transmission system operators may apply remaining physical transmission capacity for balancing purposed after the intra-day market is closed. The most efficient balancing may be reached by applying a common merit order list for the transmission system operators to bid for balancing purposes.

9.6 Challenges for the PSA Approach

The evaluation of security is essential because in the future the effects of a power system breakdown will span larger areas due the more integrated grids. New technologies and variable generation will change the dynamic behaviour of the power system. These issues may introduce a decrease in the transmission capacity delivered to the market when transmission system operators ensure the secure operation of the grid because they are uncertain of all risks the power system may have. The decreasing transmission capacity has adverse effects on the integrated electricity markets, where more congestion appears due to the more scare transmission capacity.

In this context, the PSA approach is a flexible framework and may be adapted to study future power systems and their security. The approach may be applied to any integrated power system or part of it to study the behaviour of the power system after the faults. The application of event and fault trees together with the dynamics of the power system makes possible to evaluate the importance of different components and their effect on system reliability and find out the effective ways to improve the grid security. The assumptions for the PSA approach have to be carefully selected to apply the approach for the integrated grids.

The PSA approach enables a more efficient utilisation of the grid, i.e. giving more transmission capacity to the market, without reducing its reliability. It can adapt to the changing power system and contribute with proper assumptions to the security of integrated power systems.

References

1. EC (2009) Regulation (EC) No 714/2009 of the European Parliament and of the Council of 13 July 2009 on conditions for access to the network for cross-border exchanges in electricity and repealing. Regulation (EC) No. 1228/2003, Official journal of the European Union, L211 (14.8.2009), pp 15–35

2. UCTE (2009) Operation handbook. http://www.entsoe.eu/index.php?id=57. Accessed 30 June 2010
3. Nordel (2007) Nordic Grid Code 2007. http://www.entsoe.eu/index.php?id=62. Accessed 13 May 2010
4. UCTE (2004) Final report of the Investigation Committee on the 28 September 2003 Blackout in Italy. http://www.entsoe.eu/index.php?id=59. Accessed 13 May 2010
5. UCTE (2007) Final report—system disturbance on 4 November 2006. http://www.entsoe.eu/index.php?id=59. Accessed 13 May 2010
6. EC (2009) 17th meeting of the European Electricity Regulatory Forum, Florence, 10, 11 Dec 2009. http://ec.europa.eu/energy/gas_electricity/forum_electricity_florence_en.htm. Accessed 30 Jun 2010

Chapter 10
Conclusions

10.1 Introduction

This chapter summarises the claim of the whole book: using a probabilistic method combined with power system dynamics can give information on the security of a power system by identifying its vulnerabilities, and by ranking them. By doing this, the significance of different components and their influence on the security of the power system are revealed. Applying this information, it is possible to eliminate those contributing factors that are dominant in order to reduce the residual risk. These results can then be used for grid planning, operation and maintenance.

This chapter also discusses the justification of the method, gives the interpretations of the results and proposes ideas to use the method in other ways than the one presented here. The limitations of the method are also presented.

The overall objective of the transmission system development is to strive for balanced system design, where the proportion of different components to the system breakdown would be balanced and no component would have a significant influence. This implies that a failure of a single component should not be responsible for a large proportion of the credible chains of events that can lead to a system breakdown. This chapter also presents some thoughts of how the results may be used for transmission grid and maintenance planning and proposes ideas to use the method in other ways.

10.2 Identification of Vulnerabilities

Using a PSA method enables getting intimate knowledge of the power system as the method forces to look at the security in a systematic way. It produces information that would be impossible to obtain with other means. Event trees provide the chains of events that lead to safe and unsafe situations. The strength of the method is the way in which it combines information on component failures, knowledge of structure of the power system, and hence provides information about system failures.

L. Haarla et al., *Transmission Grid Security*, Power Systems,
DOI: 10.1007/978-0-85729-145-5_10, © Springer-Verlag London Limited 2011

The PSA method systematically identifies different ways of how things can go wrong. Despite the uncertainties on the model and on component failure data, the results of the method can suggest ways of improving the system. The most important minimal cut sets and ranking lists of components are those where the improvements benefit most the whole system. It is the system and its state that matters, not the state of a single component. The meshed structure of the transmission grid, of course, is such that a failure of a single component should not be a risk. The transmission grid is meshed, but not all the components of the grid are duplicated. For example, circuit breakers are not duplicated.

The PSA method can bring risk-informed thinking into the asset management of the power systems because the method identifies the relative importance of substation components and thus helps the transmission system operator's personnel focusing on maintenance operations to the most important components when it comes to reliability. The PSA method makes it possible to identify changes in asset management that improve system reliability.

The PSA method can reveal the impact of any change in the power system or in its operation and maintenance practices on the system security. By doing sensitivity analyses, it is possible to get information on how changes in a system affect the security. For example, it is possible to evaluate how the changes in regular test intervals affect the system security.

Examples of Using the Results

The results of the first analysis for the Finnish 400 kV grid (only with a light power flow) revealed that some telecommunication channel between the distance relays at different line ends were ranked high as the contributing factors of a system breakdown. After that, the 400 kV line protection systems were improved and now they have all duplicated telecommunication channels in addition to duplicated relays.

...In 2005, the method was used by the Finnish transmission system operator Fingrid for estimating the effects of the relay test interval on security. After the analyses, the test interval of microprocessor relays was increased from 1 to 2 years.

...A study on the variation of operational reliability during different power flows was made in 2009 [1]. According to the results, the system is more unsecure when power is transmitted from Finland to Sweden than with the opposite power flow direction. This is because short circuits, which are not correctly tripped, cause more probably severe consequences in a heavy export situation than in a heavy import situation. The results also show that a circuit breaker failure after a line fault is the most probable cause of a system breakdown in summer conditions and a failure of a busbar protection relay after a busbar fault in winter conditions.

10.3 Justification of the PSA Method

The PSA analysis is risk estimation rather than calculation of the system failure rate. The method was first developed for analysing the safety functions of nuclear power plants. The power system operations after faults are to some extent similar to the safety actions of a nuclear power plant. The method helps to predict rare events and rank them. Alone, any rare event has a low probability, but ranking them helps to identify the vulnerabilities of the system. The method gives information about the components and parameters that are modelled. This is always the case with modelling, those components and parameters, which are not modelled, are lost in risk assessment.

The PSA model can be applied to a grid of real size; it gives results both at component and system level and ranks the components according to their importance. It would be easier to create accurate and detailed models for the reduced size grids, but these models often have more academic than practical value. In this book, the method was applied to estimate the reliability of the Finnish 400 kV transmission grid.

The power system is a non-linear system, where similar failure sequences in different locations and at different power flows can lead to different system states and contributions to the system security. Thus, the dynamic simulations of the substation failures are an essential part of the PSA method. Combining the substation post-fault events with grid dynamic simulations, one can get the importance measures for different power system post-fault states. When the reliability of the meshed transmission grid is under discussion, it is the system state after failure that is more important than the unavailability of a single substation component, a bay or a line.

The PSA model is detailed enough to give information about the impact of different substation components on the grid security. The PSA model can be built by using commercially available reliability and power system dynamic simulation software.

Building the reliability model for substations, and executing dynamic simulations for different power flows, requires expertise and a lot of work. After the model has been built, the information collected can be beneficial since creating the PSA model brings in-depth knowledge of the system. The PSA model is not a black box model because it reveals the interrelations of subsystems and components.

10.4 Interpretation of the Results

It is important to recognise the properties of a probabilistic approach. The probability indicates the degree of uncertainty and results like *once in 9 years* needs to be understood as a rational belief based on a certain case and certain assumptions instead of a scientific fact that can be proved. This probability model connects the evidence of the component reliability to the breakdown probability of a transmission system in a rational way.

The model gives information about the upper-level, i.e. the transmission grid, reliability by using the reliability of the lower-level components. There is no data available on the system breakdown, but there is a lot of information about the failures of different components. The grid-level failure is a function of the structural function of the system and the reliability of the system components. The important results derived from the PSA approach are failure sequences that contribute to the system breakdown, the importance values and ranking of different components, and the indicators for the system breakdown. Thus, the real result of the PSA approach is the knowledge of the system characteristics, not the exact numerical values.

The applicability of the results depends on the structure of the model, i.e. how well it represents the reality and the quality of the fault and failure statistics available. The model of the power system is usually, if not always, incomplete compared to the real power system. If the component failure statistics are made for purposes other than reliability analysis, they need to be interpreted. For example, the failure statistics are often intended for maintenance purposes. If they are used for a model of this kind, it is necessary to read every failure report and conclude from them whether the failure reported is a failure according to the PSA model or not. The same principle is valid for dynamic simulations; here, a generator trips only if it loses the synchronous operation. In reality, a generator can trip due to low voltages even though it is dynamically stable.

10.5 Possible Ways to Use PSA

The reliability centred maintenance on the system level is not possible without a system level model, for example the PSA method presented in this book, otherwise it is at component level only. The PSA method enables risk awareness in the asset management since it gives the connections between the system breakdown risk, grid faults and failures of single components. It brings the quantitative element into reliability analysis and helps to rank the substation components in different substations for the system security.

The book presents a PSA model for line faults only. Busbar faults are less frequent, but they may be more severe. The model for busbar shunt faults can be created in a way similar to the model presented in the book. The initiating event frequency for substation faults depends on the number of devices at the substation and if the substation is located outdoors or if it is a gas-insulated substation.

With wide-area measurement systems, and system level controls based on them, the power system becomes more dependent on complicated controls and their planned actions. Therefore, the failures of special protection schemes and their impact on the system security could be analysed, too.

In some cases, the component failure statistics are not available, and therefore, a quantitative modelling seems to be impossible. However, that is not necessarily the case. It is possible to use the data received with expert judgments and get information about the structural properties of the power system from diagrams and pictures. In this way, one can also find important failure combinations. There also exist methods for estimating the structural importance of the components [2]. These methods do not need any failure data.

If the fault frequencies and typical power flow patterns vary from season to season, the analysis should be repeated using relevant fault frequencies and flow patterns for each season. One way to use the model would be to estimate how often the grid ends up in different power system states during a certain moment with a given power flow and grid connection, initiating event frequency and weather forecasts (e.g. the probability of lightning, ambient temperature). This analysis can be useful during the planning stage of line or busbar outage, for example. The risk of the power system breakdown during different power flows can be compared.

The method can be used for the transmission grid planning. Different substation locations, busbar arrangements, and protection systems can be compared from the reliability point of view. This would help in the comparison of alternative investments plans.

The model presented here is for longer chains of events and more severe faults exceeding typical contingencies. To compliment the information received with the PSA model it would be useful to do a contingency analysis, systematically calculating all the credible faults and their consequences. Especially with large interconnected grids, systematic analysis can reveal the non-obvious combinations of maintenance outages and faults.

The European legal framework requires that transmission system operators shall make available to market participants the maximum transmission capacity on the interconnections and the transmission networks affecting cross-border flows. However, this requirement assumes that the secure grid operation is to be maintained. The PSA method may be applied to analyse the security of the transmission grid before and after the maximisation of transmission capacity to ensure that the system security has been maintained.

In the future, transmission grids have to adapt more variable generation with increased power flows across the transmission grids. There will be more parties acting at the grid with increasing interdependency between the grids. This change will require advanced tools for analysing the reliability of the transmission grid. The PSA method may be a tool for comparing reliability aspects of different choices for the future transmission grid.

10.6 Final Remarks

There has been little analytic or simulation work on the transmission system security. The reasons are the complexity of the problem, the difficulty of creating proper models and getting relevant data for the models. The method developed in the book is an approach to treat the security analytically and based on the knowledge of the system. The method takes into account the effect of the following issues in the matter of the system security:

- the frequency of line faults,
- fault location at the line,
- different substation arrangements,
- the failure rates of substation components,
- dynamic behaviour of the power system after different contingencies and
- reaching the remote backup distance protection.

The mechanisms that lead to a power system breakdown are diverse in different parts of the transmission grid. Thus, quantitative analysis is needed in order to correctly estimate the contribution of different fault locations and different grid components to the system breakdown.

When a transmission system operator has a tool that can really estimate quantitatively the reliability of the grid after grid faults, it can be used for several different purposes. It is possible, for example, to calculate the probability of a system breakdown during a planned outage with different grid connections and with different power flows and then decide the connection and maximum transmission capacity for that outage. It is possible to compare different substation arrangements when planning a new substation. The aim is to introduce a practical, rather than a purely theoretical, model.

The book presents a PSA-based reliability analysis of a transmission grid with some assumptions. In this way, an understanding of the system security after grid faults can be gained. One has to bear in mind that an analysis of this kind does not give a comprehensive view of all possible transmission system risks, because not all initiating events or failure possibilities are modelled. Nevertheless, the method used in the book can also be used for risk, as well as reliability, analysis.

The PSA approach enables revealing more and less probable failures of the substation actions after grid faults. It focuses on the chains of events, which can be the results of combinations of different device failures, faults and operation stages.

Since there can be an infinite number of different faults, a selection of the cases for further analysis should be made. The substation faults can cause faults, where several components trip; therefore, the initiating event frequency and different consequences of them should be analysed.

Using event and fault trees differs from random sampling methods, such as Monte Carlo simulations, because it gives the causes and frequencies of events and reveals vulnerable components, not just only consequences. Simulations based on random sampling can give results and probabilistic indexes, but the system behaviour often remains hidden. The PSA method can give connections between the causes and consequences, which is invaluable for the grid planning engineer or system operator.

Markov models would additionally give the duration of different component states, failed or healthy. Calculating the Markov model states would be laborious for a real grid with hundreds or thousands of components. If there were, say 50 lines and for each line 1–3 event trees and ten fault trees, each fault tree having five different basic events, there would be $50 \times 10 \times 5 = 2500$ basic events, each of them having two states. The combination of states would then be 2^{2500}, the consequences of which would have to be analysed. It could be possible to eliminate those states that do not need any further analysis, but even that would be a laborious task. In the PSA method, the failure rates or constant unavailability values, rather than component states, are used, thus decreasing the possible combinations to be studied for the system security.

References

1. Lamponen J, Haarla L (2009) Variation of operational reliability of power transmission system. In: CIGRE symposium in Guilin City, China on operation and development of power systems in the new context, Oct 2009
2. Myötyri E (2003) Measures for structural properties of systems. Master's thesis. Helsinki University of Technology, Espoo, Finland

Index